THE DAISY SUTRA

THE DAISY SUTRA

conversations with my dog

HELEN WEAVER

Illustrated by
Alan McKnight

BUDDHA ROCK PRESS
WOODSTOCK, NEW YORK
2001

Grateful acknowledgement is made to Penelope Smith and Pegasus Publlications for permission to quote from the audiotape *Animal Death: A Spiritual Journey*, ©1991 Penelope Smith.

The drawing on page 162 is from a photo by Nicholas De Sciose.

This is a true story, but the names of some people have been changed to protect their privacy.

Publisher's Cataloging in Publication

Weaver, Helen, 1931—
The daisy sutra: conversations with my dog / Helen Weaver.
p. cm.
LCCN 00-091946
ISBN 0-9700502-8-3
Includes bibliographical references.
1. Human-animal communication. 2.Pets—behavior—Anecdotes. 3. Animal Communication—Anecdotes
I. McKnight, Alan, ill. II. Title
SF412.5.W43 591.59

Printed and bound in Canada
First Edition

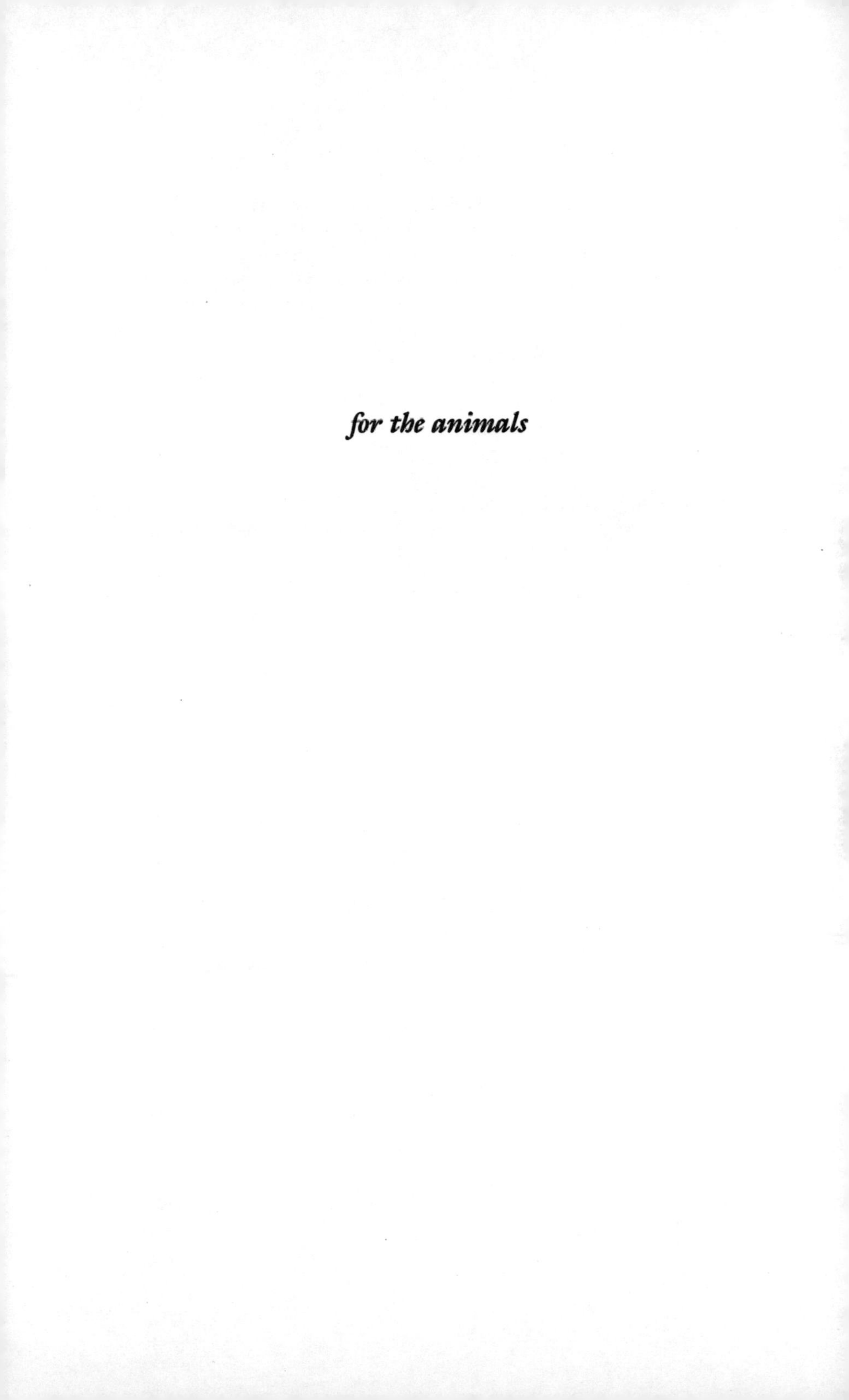

for the animals

Contents

Illustrations

New Milford

Woodstock

Beyond

Interview with an Animal Communicator

Epilogue

A Word to the Skeptical Reader

About This Book

A book has a life of its own. This one started as a memoir of my dog, Daisy. But on the day I picked up the phone and called an animal communicator, the project took on another dimension. What started as the story of one dog's life became an exploration of the possibility of talking to animals, all animals, and even, with the help of these gifted intermediaries, of hearing what the animals have to say—in other words, of dialogue.

Obviously, not everyone is ready to go along with this idea. A scientist's daughter, I was a skeptic once myself, and I have a lot of respect for people who doubt and consider and weigh the evidence. Writing this book inevitably reminded me of my own journey from skepticism to belief, and I recall that process in a little piece I call *A Word to the Skeptical Reader*.

If you tend to be skeptical, you might want to start with that, or with *Interview with an Animal Communicator*, before turning to *The Daisy Sutra*.

If you would like to learn more about the rapidly expanding field of animal communication, you will find a list of recommended books, tapes, and animal communicators in *Resources*.

When I began writing about Daisy in my journal toward the end of her life, I had no idea that my scribbled notes would become a book. It's been an amazing process and although I've put in many obsessive hours, there's a sense in which the book has written itself. Help and support has come from all four directions, and from many hearts and hands. I honor and acknowledge them all.

I'm grateful to my parents for getting me my first dog, and for their perfect example of kindness to animals. Books about animals, especially the Pooh books and *The Wind in the Willows*, were family favorites. My thanks go to A. A. Milne, Kenneth Grahame, Anna Sewell, Albert Payson Terhune, Marjorie Kinnan Rawlings, and all the other writers who nurtured my love of animals by sharing their own.

My love and thanks to all my animal companions: Brownie, Misty, Petra, Pluto, Max, and Daisy, each a unique personality, a teacher, and a friend. Without animals, how barren our lives would be!

And without the insistence of the unforgettable woman I call "Dolly," I might never have thought of adopting a dog and might never have known Daisy. I am eternally in her debt.

To Shellie David, who first told me about animal communication, and to the animal communicators—Gail De Sciose, Ginny Debbink, and Karen Beth—my gratitude is profound. Without them, this book would not be possible. Gail agreed to let me interview her in the midst of a dental crisis and has given freely of her time and assistance. I thank her from the bottom of my heart.

I also acknowledge my debt to Gail's teacher, Penelope Smith, a pioneer who has led the way in animal communication for over twenty years. Her dedication and courage are helping us all to learn what Native Americans have always known: that animals are our teachers and our guides.

I want to thank all the friends who read early versions of this book and gave me much needed encouragement and often, valuable suggestions: Linda Baker Abrahamsen, Sarvananda Bluestone, Annie Weaver Buck, Bhabi Burke, Sam Chetta, David DePorte, Melissa Weaver Dunning, Michael Esposito, Gerald Fabian, David Goldbeck, Michael Green, Elaine Grinnell, Hariet Hunter, Hatti Iles, Michael Korda, Janaan Moncure, Marcia Newfield, Bill Riordan, Carrie Schanze, Ellen Shapiro, Martha McGehee Teck, Dan Wakefield, Sally Weaver, Susun Weed, and Clarisse Zielke.

I am grateful to the skeptics in my life, who keep us believers on our toes, especially Marianne Means, whose willingness to listen to new ideas is, I hope, a sign of the times.

I'm much obliged to computer maven Jon Delson, who rescued Daisy when she was lost in cyberspace and has provided ongoing technical support.

I feel very lucky to have connected with two such talented artists as Alan McKnight and Cheryl Taylor. Alan's drawings and Cheryl's design work have added beauty to the book, and they were both so patient with me that I suspect them of having transcended their egos.

As for my two extraordinary editors, Miriam Berg and Gale McGovern, I shudder to think of what this book would be without their painstaking help and inspired guidance. Working with them has been a privilege and a pleasure, and my debt to them is enormous. Any faults and errors that remain are mine and mine alone.

Finally, a deep bow to the spirit of Daisy, my co-author and my muse. May our book serve the highest good of all beings and help in the healing of the planet.

Woodstock, New York
Summer, 2000

But ask now the beasts, and they shall teach thee;
and the fowls of the air, and they shall tell thee;
or speak to the Earth, and it shall teach thee;
and the fishes of the sea shall declare unto thee.

Job 12: vii-x

The practically lost art of listening is the nearest of all arts to Eternity.

William Butler Yeats

The more people who know we are not just dumb animals, the better.

Daisy Weaver

Prologue

When I got back from Saugerties Animal Hospital, I took Daisy's old red collar and put it around my neck. It was a tight fit, and when I walked, the license and rabies vaccination tags clinked together with that familiar sound. It was eerie. So I took it off and got a pair of pliers and pried the tags off the metal ring on the collar. I found an old chain in my dresser and threaded it through just the license tag. It's a golden disc with eight scallops that look like petals—very appropriate for a dog named Daisy who loved the sun. I put the chain around my neck. I needed to have something that had touched her touching me.

Down in the den I found some pictures of Daisy and put them all around the house: on the altar in the meditation room, on my desk, on the dining room table. I remembered how when Miriam's cat Honeypie died suddenly of feline leukemia, her picture appeared on Miriam's refrigerator. I understood the impulse. I needed to look at Daisy, and I needed other people to see how beautiful she was and to remember her.

And now I need to tell her story, to celebrate her life. Because I promised her I would. Because having lost her, I need to save my memories. But also because Daisy's story is part of a bigger story, a story about love and listening and spirit, a story that is unfolding all around us as we learn more about our animal friends. I need to tell you what I learned from this little dog who was, and still is, my teacher.

New Milford

Daisy came into my life in October of 1983. Those were dark days for me. I had given up my home in Woodstock, New York, and moved into my mother's house in New Milford, Connecticut, to take care of her in her old age.

A conservative New England town whose green, featuring a World War II tank and a gazebo, once appeared on the cover of *The Saturday Evening Post*, New Milford was not my town. My town was Woodstock, whose postage-stamp-size green features a bearded tarot reader wearing rainbow-hued patches and a yarmulke, street people, and flowers. My time spent in Connecticut felt like exile, especially in the beginning, before I made some friends who reminded me of home.

The morning after moving day I woke up with a sinking feeling. I was sure I had just made the biggest mistake of my life. *Oh my God—what have I done?* I tried to swallow my panic. I repeated the mantra that had got me through packing: *I'll try it for three years.*

My mother's mother had lived to be 93. Mother was 87 when I moved in. If I had known then that she would live to be 100, I'm not sure I would have made it. In one piece. Although the open-endedness of it—the sense of suspended animation the caregiver lives with when she puts her own life aside, the fear that this holding pattern will go on forever—was bad enough.

Nobody in their right mind would do this, I told myself. And on my better days, *You don't have to be crazy to work here—but it helps.*

Here I was, at age fifty, complete with salt and pepper hair and hot flashes, back on my parents' turf. This was their dream house, not mine: the elegant ranch they had built on the top of a hill overlooking the town, designed by them to be first their summer home and ultimately their year-round residence after my father had retired from his job in the city. I had never felt at home there. I have photos of myself taken in the fifties in Mother's rose garden, dressed all in black, hair dykily short, wearing sunglasses (we called them "shades") and a blank stare, with equally alienated friends of both sexes. My beatnik period: a fish out of water if there ever was one.

New Milford

Before the house was built, my parents had put up a garden house, a place to store the Gravely tractor and garden tools, with a tiny porch and a built-in barbecue, where they could drive up from Westchester and have picnics while staking out the land. The garden house had electricity but no heat or running water. It had long since been abandoned when my parents got too old to work in the garden. This was more my style, and when I discovered that the fieldstone chimney that served the barbecue contained an inside flue as well, I blessed my father's foresight, hooked up my wood stove, insulated the whole place, and claimed the cabin as my own.

I have something in me that is drawn to solitary confinement. I had always had a bizarre fantasy that I could stand to live in prison as long as I was allowed to paint the floor of my cell white. Why white floors? Because ever since I'd seen the white-washed streets of Molivos on the Greek island of Lesvos, and then the poet Frank O'Hara's loft in lower Manhattan, white floors meant freedom, meant catching all the available light, meant living in the clouds, or in the purity of new fallen snow, or of the empty page. I painted the floor of the cabin white.

But although the cabin provided me with a rustic privacy I needed, I was still a fish out of water. The Berkshires were not the Catskills. The land was tamer, more clipped and combed. The people were more conservative. This was New England, after all. It definitely wasn't Woodstock, with its racy blend of artists, musicians, and aging hippies (among whom I numbered myself), where it was no big deal if a stranger smiled at you on the street.

This was not the first time I had taken care of my mother. One day toward the end of my father's life, Mother had gotten up on a stool to fish something off the top shelf of her bedroom closet. She fell and broke her arm and in the panic of her fall, she lost all muscle control. Fortunately, I was there visiting at the time, and after I called the ambulance, I cleaned her up.

A fastidious woman, Mother was humiliated. "I'm so sorry you have to do this!"

I laughed. "Mother, how many times have you done this for me?" It felt good to be able to take care of her for a change. As the black sheep of the family, it felt good to be useful.

My father wasn't in great shape either, so I stayed on to take care of them both while my mother's arm healed. Caregiving was satisfying—in small doses. Now I began to see that I had bitten off more than I could chew.

The good news was that I was not my mother's only caregiver. My father, a good provider, had left enough money for me to hire helpers. At first, when Mother took to her bed with a severe case of flu that almost killed her, I tried to do everything myself, but eventually she had to have nurses around the clock. As she improved I switched over to live-in companions. After a few false starts, I found Dolly.

New Milford

Of Japanese-Korean parentage, Dolly was a small woman of indeterminate age, still pretty, a tough cookie whose charm and energy were a breath of fresh air in our somewhat somber menage.

Dolly had been in our employ less than a week when she looked around the living room, hands on her tiny brown-trousered hips, and uttered the nine words that changed my life: "House too big. You need dog. I take care!"

I called the local Animal Welfare Society. Back then they didn't have a permanent home; they were a referral service for homeless animals, who were farmed out temporarily to various vets and animal lovers in the area. They had two dogs who were up for adoption. The first dog, on a long lead in someone's yard, wrapped itself around a tree within seconds of my arrival, barking hysterically. Regretfully, I had to decline.

The second dog was being kept in a kennel at New Milford Animal Hospital on Route 7. I drove down to check her out. As I sat on a plastic chair in the waiting room Daisy was led out on leash. She was a medium-sized black dog with white socks and belly and a white collar around her neck—hence the name the employees had given her. She came right over to me, wagging her tail in a decorous manner, not at all aggressive, but friendly and alert.

Daisy had been described as a shepherd mix, but there was as much beagle as shepherd and a hint of collie in her longish nose. I remember that I wasn't particularly impressed

by her from the side, but once she turned her head toward me and I looked into her beautiful, expressive brown eyes, it was all over. I reached down to pet her, and only then did I perceive that she was trembling violently. Her placid, polite manner was a big act: this dog was terrified. Later I learned how much Daisy hated being in a cage. At that moment, when my hand touched her vibrating back, I understood what she was telling me with every cell in her 35-pound body: *Get me out of here!* And I thought to myself, What intelligence! What acting ability!

Much as I was tempted to take her home with me that day, I thought Mother and Dolly should have a chance to meet her, too. An expedition was arranged for the next day.

Heroically, flanked by Dolly and me and manipulating her quad cane, Mother managed to negotiate the steep stone steps to the vet's waiting room. Daisy was led out again and went right to Mother, as if she knew that she was in charge. She waved her tail in a ladylike fashion, greeting Dolly and me in turn.

"So beautiful face!" was Dolly's comment, and Mother and I agreed. "She good dog. You wanna come home with Mama?" Once again Daisy was trembling from head to tail under her calm demeanor. I signed the papers and paid the modest spaying fee, and off we went.

Daisy's first act upon arrival at our house was predictable, given the stress she had been under for the past few weeks.

"Oh—she do poo poo!" Dolly announced superfluously as the old green carpet in the living room received Daisy's involuntary benediction. Of course we all forgave her.

Daisy had lucked out bigtime: ten acres of her own, and because we lived at the end of a long driveway, she never had to be tied up like some country dogs.

Remembering how a previous dog, Pluto, had got Mother in trouble by wandering off in search of amorous adventure, I decided to teach Daisy to remain within the borders of our land. I had Edward, our handyman, cut a path around the property just inside the boundary line. I walked this path with Daisy several times, telling her solemnly with voice, eyes, and gestures, "This is our land" and "This is not."

Daisy learned her lesson so well that later, when I wanted her to accompany me on longer walks to the mailbox and beyond, she refused. She would plant her bottom on the driveway at the exact spot where our property legally ended and look at me as if to say, "That's not allowed—at least, that's what you *told* me." Eventually she did agree to walk with me, but I don't think she ever quite forgave me for changing the rules on her.

She loved our walks, of course. What dog doesn't? A dog loves two things above all else: her person, and being outdoors. So it stands to reason that when she is outdoors with her person, it's the best of all possible worlds. If her person is like me, i.e., not very good about remembering to get enough exercise, then a dog is a very good influence indeed.

When she saw me pick up her leash, Daisy would get up on her hind legs and dance and sing. The word "beagle" comes from the Old French *beegueule*, open throat, and beagles are the singers.

She had to be on leash until we got well past Duke, the German shepherd who guarded our neighbor's estate. Every once in a while their gate was left open, and one of those times Daisy scampered gaily into Duke's domain. He came roaring out of his doghouse, and we both got out alive, but it was scary. Another time Daisy tried to attack a pair of horses on Legion Road, and again we all narrowly escaped an accident.

No, Daisy was not trained, let alone "at voice command." Her only trick was giving you her paw to shake, and she only did that when she felt like it. So the leash was important.

But at the end of our walk, having survived Duke and reached the top of our mowed field, we would squeeze through the gap in the split rail fence, and I would take her off leash. Daisy would plummet down the hill between the big oaks. She had a kind of diagonal tilt to her low-slung plunge that always made me think the word "skedaddle." But every fifteen or twenty feet she would stop and look back over her shoulder to make sure I was still there behind her. Then she would race off again, wriggling all over with the double joy of togetherness and motion.

I always thought that diagonal tilt was peculiar to Daisy. But on rereading Thomas Mann's *A Man and His Dog*, I learned that this lopsided lope, in which the hind legs move not directly behind the front legs but somewhat to the side, is the common habit of dogs.

Daisy was *fast*. She could nail a squirrel at fifty paces, and those long-tailed gray squirrels of Litchfield County were no slouches themselves—why, they could practically fly from tree to tree. There are two schools of thought on squirrels, as there are on deer, or any other species that is pesky but cute. There are those who feed their squirrels along with their birds, and those who view the charming rodents as clever thieves to be outwitted by better bird feeders. (Lots of luck!)

I had hung a supposedly squirrel-proof bird feeder outside Mother's bedroom window where she could watch her beloved chickadees, tufted titmice, juncos, and the occasional cardinal while waiting to be served breakfast or during the commercials of *Days of Our Lives*. We all watched with mingled annoyance and admiration as the squirrels gradually mastered the most sophisticated bird feeder Village Hardware could provide. You had to respect their persistence and cunning. There were times, in the early stages of their research, when they would sit on the ground staring up at the feeder, and you could swear they were making complex mathematical calculations in their little fur heads.

Although we wouldn't let Edward shoot them, Mother and I more or less belonged to the anti-squirrel camp, which made it easier to forgive Daisy, the mighty hunter, who saw them as dinner.

Daisy's appetite was legendary, and to say that her tastes were catholic is putting it mildly. After dog food, people food, and live game, paper products were her special favorites: paper towels, toilet paper, and Kleenex were some of the least perverse of the items she considered edible. She ate half a postcard from my niece Annie, including the stamp, and once, an employee's paycheck. I understand the phenomenon—papyrophagia? papyrophilia?—is not uncommon in dogs, but neither is it very well understood. At any rate, no wastebasket was safe from her depredations; each one had to be elevated, kept in a closet, or the step-on variety. A paper towel, for example, would travel through her digestive tract and emerge virtually unchanged. When the snow melted in the spring I would find these interesting white sculptures on the lawn.

Food could by no means be left unattended. I once saw Daisy wolf down in seconds a package of frozen hotdogs that had been left on the kitchen counter. Much later, after Mother died and the house went on the market, the building inspector left half a ham sandwich in his car with the back end open. Daisy didn't even bother to unwrap it. Fortunately, the man had a sense of humor.

A healthy dog is always hungry, we're told, but Daisy's hunger had a hysterical edge to it that made me wonder about her past. The people at Animal Welfare were not sure of her previous history. She had been picked up on the street, and rumor had it that her owner was a young man who had died of cancer. Whatever the truth of this, mealtime was the high point of her day. You had the feeling that no matter how much you gave her, it was never enough, and that hunting, besides being her beagle heritage, was her way of supplementing the stingy rations her people fed her.

Irma, who succeeded Dolly as a live-in companion, doted on Daisy, but she also had a soft spot for the squirrels. Whenever she let Daisy out Irma would yell, "Up the trees!" to alert the squirrels that their nemesis was at large. Daisy was a good hunter: she ate every animal she caught, or at least every part she considered edible; the rest she rolled in or buried against hard times.

Still, it was hard to forgive her for the chipmunks, the baby birds, the possum. And as for the beautiful white rabbit she caught and killed while Irma and I watched, horrified and screaming, from the back door—we didn't speak to her for two days after that. Later that morning I saw Daisy strut across the lawn with a long black scarf in her mouth, heading for the woods. That afternoon I found the black scarf on the lawn and recognized it as all that was left of the rabbit. She had recycled the rest.

But that was later, after Dolly had left us.

Meanwhile, Dolly and Daisy were a comic team that helped Mother and me and my niece Annie, who lived with us for a while, from getting too down in the dumps. Unbeknownst to any of us, Mother was suffering from the depression that is so common in the elderly and so often goes undiagnosed. I missed my friends and my life in Woodstock and was going through menopause. Annie had just dropped out of art school and was at loose ends. Dolly was right: we needed dog.

Dolly's job description included grocery shopping. The only trouble was, she didn't have a driver's license. She knew how to drive and she promised to get one right away. What I did not realize when I hired her was that she couldn't read or write English. Dolly was a good driver and as smart as they come, but how was she going to pass the written part of the test?

Dolly said, "No problem. Buddha help me, I pass. You see!"

Dolly was a recent convert to the dynamic Soka Gakkai school of Nichiren Buddhism, a sect founded by a thirteenth-century Japanese priest and firebrand. Adherents have a daily practice of chanting *Nam-myoho-renge-kyo* ("Homage to the mystic law of *The Lotus Sutra*"). Dolly set up a little portable shrine in the room off the kitchen which served as her bedroom. She even had Annie and me chanting for a while ("You see, you get benefit!"), until she took us to a meeting in Hartford that felt more like a pep rally or an Amway meeting than a spiritual gathering. Members told how they had received enough money to buy a new car by chanting *The Lotus Sutra*. I spent the evening fending off the advances of a head honcho woman who wanted to sign me up for only twenty bucks.

Raised as a Congregationalist Christian by a scientist father and a skeptical mother, I was drawn to Buddhism as I aged. My own practice, if you can call it that, is eclectic, to say the least. Alan Watts, the early popularizer of Buddhism in the West, once asked the great mythologist, Joseph Campbell, "What's your yoga?" (meaning, what's your spiritual practice?). Campbell summed it up in two words: "I underline." Although I've recently taken to highlighting, I'm in Campbell's camp.

But sure enough, when the time came for Dolly to take the written test, the Buddha told her the right answers and she passed with a near-perfect score.

Or maybe she threatened the guy who gave her the test at Motor Vehicles with bodily harm. Dolly was tiny, but she was strong. Before she came to work for us, she worked the 3 to 11 shift at a hospital in Waterbury in a notoriously bad neighborhood. As she was leaving work one night around midnight, two big men moved in on her in the parking lot.

Dolly decked one and the other one took to his heels.

Dolly knew shiatsu as well as karate, and I had her work on me a few times. Her arms and hands were like steel, and her version was so intense that I would scream and pound the living room floor like a wrestler on TV. It must have looked as if Dolly was trying to kill me, because Daisy would come over and try to get between us. I had to explain to her that I really liked it.

Dolly had a husband and two little girls back in Arizona. Her husband was a hairdresser. When Mother asked why she didn't live with them Dolly said, "Hmmm, Mrs. Weaver, I marry him, I not know, he is be—a homosex!" Mother laughed so hard I thought she was going to have a stroke. Whenever she told the story, Mother would giggle helplessly, dabbing at her eyes with a Kleenex from the box next to her chair. She was not without sympathy for this woman adrift with her broken English. When Dolly was out of earshot Mother would lean forward, shake her head, and whisper, "Pathetic!"

As a young bride in Pasadena, my mother had tutored the Japanese wives of my father's fellow professors in English. She offered to do the same for Dolly and her offer was enthusiastically accepted. We managed to unearth the dog-eared copy of *English for Foreigners* (Sara R. O'Brien,1909) Mary Weaver had used in the 1920s. It's a small treasure of a book with the image of the Statue of Liberty stamped on its cloth cover in fading red ink and an early photograph of "The

Capitol At Washington" facing the title page. The lessons began, but progress was slow. Dolly's great gift was for comedy, and we all loved her for making us laugh.

One day Dolly, Annie, and I were sitting around Mother's chair in the living room when Daisy, for some unknown reason, began humping Mother's leg. Dolly looked at the dog and announced in that emphatic way of hers, "She *gay*!" Annie giggled, I howled, and Mother rolled her eyes heavenward, shaking silently, while Dolly basked in our delight. If you say that phrase with the proper emphasis, it sounds very Japanese. "She *gay*!" became as much a part of the Weaver family legacy as the good silver or the Thanksgiving china.

I should perhaps explain that the Weaver family consisted at that time of my widowed mother, whose eighty-some years sat on her with a grace that gave new meaning to the phrase "well preserved" ("No stress," I was wont to observe, with a bitterness only partly feigned); my brother Warren, eight years my senior and my only sibling; his former wife, the mother of his children; his present wife; and his four blonde daughters, each more beautiful and talented than the last, of whom Annie was the youngest. Annie was a special favorite of my mother's and it was generally agreed that the particular quality of her beauty—a certain old fashioned English purity reminiscent of Austen and Trollope—was the most like Mother's of any of the girls.

(And, of course, my mateless, childless self, a big disappointment to my mother, who believed that having children was every woman's reason for being.)

Annie's arrival had posed a problem, because Annie was a cat person, and she planned to bring her cat Mooey. Now, Daisy chased our neighbor's cat whenever the cat came into our yard. As a further complication, Mother and I were both allergic to cats, so Mooey would have to be confined to Annie's room. Edward rigged up something on one of the guest room windows so Mooey could go outdoors without having to go through the house. But the cat was bound to get out of Annie's room now and then, or meet Daisy outdoors, and what if Daisy decided that he was dinner?

Daisy adored Annie, and forgave her for not being a dog person. One day not long before Annie moved in, I had a talk with Daisy.

I looked into her eyes and said, "Annie is coming to live with us for a while, and she is bringing her cat. Annie's cat is special. She's a member of the family. You are not to chase Annie's cat! You are to welcome her to this house and treat her kindly."

On the day Annie arrived, as her car appeared at the top of the driveway, I told Daisy, "Now, remember our little talk. Annie is here with Mooey, and you are to make them *both* welcome."

Annie parked her car, laden with everything she owned that wasn't in storage, and got out. As we had planned, she let Mooey out of his carrier as Daisy and I came out the front

door. When he saw Daisy, Mooey arched his back and hissed. But Daisy walked calmly over to the hissing cat, waving her tail gracefully from side to side like a perfect hostess, and touched noses with the doubled-up cat. All that was needed to make this scene complete was a little ruffled apron on Daisy. Annie and I just looked at each other.

Annie lived with us for a year. Daisy never chased her cat.

🐾 I'm not sure I would have thought of having that little talk with Daisy if it had not been for my adventures with animals in the wild.

Shortly after I moved to Woodstock from New York City, I bought five acres of raw land a few miles from town and put a house on it. But, inspired by my parents' example, I first built a small cabin in the woods, a rustic affair with no electricity or running water.

Although I had been camping a few times with friends, I

had never slept alone in the woods. I was still living in an apartment in town. But after my builder finished the cabin I put my sleeping bag, chamberpot, kerosene stove, and a few pots and pans in my VW Bug and drove out to my land.

I arrived after dark. Just as I set foot on the path I had cleared to the cabin, out of the night there came a blood-curdling sound. It was loud and anguished, like the cry of a soul in pain. I'd never heard anything like it in my life. I told myself it must be some wild creature, but it sounded almost like a human infant. I wasn't about to turn back, so I silently addressed whomever it was that was making the sound:

"Hello, whoever you are. I am Helen. I have bought this land, but I know that it doesn't really belong to me. It's really more yours than mine. Whoever you are, you belong here. I realize that I am invading your territory. I acknowledge your prior claim and I apologize for the invasion. I hope we will be able to share the land as friends. You have every reason to be upset at my unexpected arrival. However, if possible, I would appreciate it if you would not make that sound all night."

I made my way down the path to the cabin. The sound continued, but at longer and longer intervals. It was still loud, but not quite as scary. I laid out my sleeping bag on a foam pad and undressed. Just as I blew out my kerosene lamp and pulled the sleeping bag up around me, the sound stopped. As I drifted off to sleep, I thanked my unknown friend for her dramatic welcome, her perfect timing, and her gift of silence. My first night alone in the woods was peaceful.

New Milford

The next day I went back to town for more supplies, returning to my land while it was still light. The moment I set foot on the path to the cabin, the sound started again. Once again, I addressed my invisible welcoming committee.

"Hello, friend. What a loud voice you have! I would love to know who you are. Maybe sometime we will meet. In the meantime, I want to thank you for your consideration in honoring my request for a quiet night. I think we're going to get along fine!"

I still had no idea what sort of animal this might be who announced my arrival with such precision and power. But I was soon to find out. For when I opened the door of my cabin and looked out the back window, there on the branch of a tree sat a small owl—a screech owl, of course—and she was looking right at me. She gave a final ear-splitting scream and then fell silent.

It was one of the great moments of my life. It felt like a visitation—an annunciation, almost. Ignorant city slicker that I was, I still knew that owls are nocturnal and that this was unusual behavior. We stared at each other for a few moments.

Finally, I took a deep breath and said, "What a magnificent owl you are, to be sure! I am deeply honored that such an august creature as yourself would deign to pay me a personal visit. When I first heard your voice, I was frightened, but now I am not, for I know that you are intelligent and kind.

"I was curious and said I would like to meet you, and

here you are! I promise that I will try to be a considerate guest in your home and to be worthy of the trust you have shown me."

Well, words to that effect. What struck me at the time was that throughout this rather flowery speech, the owl continued to stare at me with her huge round eyes as if she were drinking in every word, and I got the feeling she approved of me. I don't know how I knew to talk to her, it just seemed like the right thing to do at the time. They were her woods. She had every right to scream at me.

The interesting thing is—she never did it again. I felt as if I had passed some sort of test. I did see her again, though, and again her visit was unexpected.

I had moved into the cabin for the summer, had set up my manual typewriter on an old oak table and only went into town a couple of times a week for fresh produce and other supplies.

Although the owl no longer screeched at me, the woods had a new surprise for me, and this one came in the middle of the night.

Around 3 A.M. I was awakened by the sound of something at the door of the cabin. It sounded like a fairly large animal hurling itself against the door, as if trying to break it down, like a cop in a movie. As I lay rigid in my sleeping bag whatever it was gave that up and began circling the cabin, thumping on the walls and rattling the windows as if trying to find a way in. I knew there were bears in these hills. Could

a bear break a window? Then the thumping stopped.

Summoning all my courage, I grabbed the flashlight I kept next to my bed and got up and looked out the window. There in the clearing in front of the cabin was a large raccoon. Not a bear, but still wild, with teeth and claws. It lumbered off into the woods, and I went back to bed.

But my nocturnal visitor was not about to give up. The raccoon must have been smelling the food inside my cabin, and it took to coming by, always around 3 A.M., and circumambulating the cabin, checking the doors and windows for a way in.

Across the road my neighbor, Michael, lived in a tipi year round, chopping wood and carrying water from a nearby stream. Michael had said that a good way of getting the animals to respect your territory was to do what dogs and wolves do, and mark your boundaries with pee. I started marking a wide circle all around the cabin and saving the chamberpot for inclement weather.

Back then I rarely had to get up in the middle of the night to answer a call of nature, but one night I did. I took my flashlight and went outside, leaving the door of the cabin ajar. I was squatting in the grass at the edge of the clearing when suddenly out of the night came a terrific whirring sound right over my head. Looking up, I saw my friend the owl. After buzzing me like a low-flying plane, she landed on the branch of a tree.

I tried to pee, but I couldn't. Every time I started to relax the owl would fly right over my head, almost touching me. After a while I gave up and went back inside and used the chamberpot instead. As soon as I got back in bed, the raccoon was at the door. It was almost as if the owl had been telling me, *Get inside*.

When I told Michael about it he said, "You left the door open. The raccoon was there, waiting to get in. A raccoon is dangerous only when cornered. If he had gone inside the cabin before you, you would have been in trouble. The owl was warning you.

"Raccoons are born thieves, and they love to steal from humans," he went on. "A raccoon will spend all night picking a lock rather than make an honest living fishing—which they're really good at."

"So, what am I going to do about this damn raccoon?"

"Get yourself a little pile of stones and put them right next to the door. The next time he comes, let him have it!" Now, Michael was a Buddhist and would never harm an animal, so I was a little shocked.

"You don't want to hurt him, just get his attention and show him who's boss."

I decided to take his advice, and assembled my arsenal of stones. My cabin had a Dutch door opening onto a tiny porch with a shed roof. I figured I could just open the top half of the door and throw a stone without giving the raccoon an opening.

That night the raccoon came as usual. I waited until he had circled the cabin, testing for openings. When I heard him out in front, I opened the top half of the door and looked out, flashlight in hand. There he was, sniffing at the ground. I picked up a stone and threw it. It was wildly off, landing ten feet behind him.

The raccoon looked at me, walked over to the stone and sniffed it, and then looked back at me as if to say, "Can't you do any better than that? In the first place, you're a lousy shot. In the second place, if you're going to throw something, at least make it interesting."

He waddled over to the tree that stood at the foot of my porch, climbed it, and sat on a branch staring at me, radiating contempt. All of a sudden my fear of the animal changed to irritation. I opened the door and stepped out onto the porch,

being careful to close it behind me. I looked the raccoon straight in the eye, took a deep breath, and said in a firm, schoolteacher voice,

"This is *my* cabin. This is *my* territory. You are not to come here in the middle of the night and wake me up! You are not to come up on this porch." (Pointing down.) "It won't do you a particle of good. Not a particle! I will put my leftover food there on the stump." (Turning my head and pointing across the clearing; the raccoon turned his head and looked where I was pointing.) "You are welcome to take it. But you are not to come here and wake me up in the middle of the night."

Throughout this speech, except when I pointed to the stump, the raccoon never took his eyes off me. Somehow I knew that he was taking it in. I even had the feeling he was flattered that I was talking to him. I paid him the compliment of believing he could understand me. I gave him the benefit of the doubt.

After that the raccoon never set foot on the porch of my cabin or bothered me in the night. The food scraps I put on the stump were always gone in the morning.

Daisy was on her best behavior with Annie's cat. With people, however, she could be very rambunctious. She had a habit that not everyone appreciated of jumping on people. Since Dolly fed her in those days, Dolly got jumped on a lot. When Daisy scratched her new oxblood loafers, Dolly was not pleased. She had paid good money for those shoes. Annie and I heard her shouting at the dog, "You painy butt! How much you got you bank account? Gimme forty bucks!"

The more Daisy loved you, the more she jumped on you. I think it was very important to her that her special people—Annie, and my new Connecticut friends Clarisse of New Milford and Bill of Brookfield—*know* that she recognized them and that they were her favorites.

When Clarisse or Bill came over for a visit, Daisy would become ecstatic, jumping on them and singing her beagle song of love. Nothing would do but they had to get down on their haunches and pet her until her passionate welcome had subsided enough for us to talk.

Clarisse, a Cajun firecracker from Louisiana with a beautiful soprano voice, would sing, "Daisy, Daisy, give me your answer true!" and Daisy would make it a duet. Then Clarisse and I would imitate Daisy's impassioned aria, and the duet would become a trio. Clarisse and I got so good at barking that we sounded like hounds baying at the moon. Daisy would lift her little snoot in the air with the wild abandon of a dog who had finally found her pack.

Bill was also a dog person, and Daisy fell in love with him on sight. A devout environmentalist and former surveyor, Bill had retired from that profession because surveying is the first stage in land development. It was part of his religion to climb a tree every day. He seldom wore shoes, which made his feet more interesting than most.

If Bill and I made the mistake of trying to have a hug *before* he acknowledged Daisy's presence, we got jumped on from behind. In fact, when *any* two or more humans hugged in her presence, Daisy jumped on them from behind, insisting on being included in the greeting. Born in the eighties, Daisy was a dog of the sixties, a love dog, and a devout believer in the group hug. Bill was a sixties person, too, and the two of them were kindred spirits. Even before Bill and Zada's dog Chipmunk died, Bill told me that Daisy was his favorite living dog.

When Daisy jumped and sang that way, it felt like she was

trying desperately to be as tall as we were, and to communicate her love. As Dolly put it so aptly, "She try wanna talk."

Dramatic Daisy! When I called her Sarah Bernhardt, I never meant to imply that her carrying on was only an act; it came straight from the heart.

Perhaps because she was an only pet and her person had no human children, Daisy acquired many names. To Dolly and Annie she was Dayje (as in *déja vu*); to Clarisse she was Daisy Mae; to Irma she was That Little Girl. And to me—well, here we get beyond the bounds of anything that could be uttered in public. And indeed, most of them surfaced after Daisy and I were living alone: Dayzickle Wayzickle; My Blackums Wackums; Pookums Do and the Don'ts (don't ask); My Little Black Angel From Heaven; Waggy Waggy Wagtail; Miss Dog; Baby Girl; Baby Dog; and the all-time favorite, Itsy Bitsy Baby Doggie, which is the first line of a song, sung to the tune of *Marching Along Together*, the doggerel lyrics of which are revealed here for the first and last time:

> Itsy Bitsy Baby Doggie,
> Itsy Bitsy Baby Dog (two! three! four!)
> Itsy Bitsy Baby Doggie,
> Itsy Bitsy Baby Dog (two! three! four!)

Have you ever looked "doggerel" up in the dictionary? Its root is a Middle English word meaning "poor, worthless" from *dogge*, dog. Middle English dates from approximately

1100 to 1500 A.D. How much have we learned in the last five hundred years?

A people dog, Daisy loved company in general, especially children, who were closer to her size, willing to play with her, and best of all, tended to spill food on the floor at mealtimes. Begging at table was not permitted in our house, but Daisy was allowed to lie quietly under the dining room table. So that's where she would be, ever on the alert, her chin resting comfortably on the cross-brace, or if my nieces Sally or Melissa were visiting with their kids, stationed directly under the youngest and sloppiest child.

The official greeter at Annie's and my annual yard sale, Daisy kept us company, entertained the children while their parents browsed, and probably added to the success of our sale. Like a cat, she always lay in the sunniest spot. Even on the few sultry days we got in western Connecticut, Daisy

would go to her favorite spot on the blacktop driveway and lie there panting and basking in the sun.

She loved to watch me work outdoors, stacking the wood Edward cut for the stove in the cabin where I slept or weeding the garden. She would wag her whole back end when she saw me go for the wheelbarrow, and race me to the woodpile. She was good company.

Her favorite spot in the house was by the door, but as a watchdog she belonged to the "Hi! This way to the silver!" school. Any stranger who came to the door was welcomed as a potential source of food, strokes, or fun.

Although she learned not to beg at table, Daisy's behavior in the kitchen was another story. Daisy reminded me of Willie Sutton. When they asked him why he robbed banks Willie Sutton said, "Because that's where the money is." Why did Daisy beg in the kitchen? Because that's where the food was, as any fool could see.

When people were cooking, snacking, or just hanging out in the kitchen, Daisy was simply incapable of suppressing her enthusiasm for the whole idea of eating. Even after her hearing started to go, the sound of plates being scraped after dinner could bring her out of a sound sleep at the other end of the house. Preparations for her daily feeding elicited a beagle bugling of operatic magnitude. "Hold onto your bridgework, Daisy," Irma would plead.

Once when my friend Harriet was visiting from Woodstock and we were in the kitchen, Daisy was being a real pest. Harriet, who had lived in London and had higher standards of pet behavior, observed rather drily, "It's vulgar." Whereupon Daisy looked at Harriet, stopped what she was doing, turned, and walked out of the room with what could only be described as offended dignity.

Daisy's ability to understand English, or at least to understand what was being said on a nonverbal level, received further confirmation. There were times when, in order to travel, I had to leave her in a kennel. The first time I called New Milford Animal Hospital about boarding her for the weekend, Daisy expressed her opinion of that plan by making a mess on the kitchen floor. After that, whenever I picked up the phone to call the kennel, she would stand in the kitchen and howl.

How many words do animals understand? It is obvious that a dog knows her name. I remember as a child sitting and having lunch with my parents on the porch of our summer cottage while our dog, Brownie, was lying half asleep in the next room.

My father was telling my mother some story about his work at the Rockefeller Foundation, and he happened to mention Brown University. We were all touched and amused when Brownie got up, walked out onto the porch, and looked up at my father expectantly.

Many years later I was living with my parents in

New Milford

Connecticut while recuperating from hepatitis. I had just come back from Greece where I had acquired not only jaundice, but a Russian blue cat named Petra. It had been a rough voyage on the *Olympia* in November, and on arriving in New Milford, Petra had come down with a bad case of colitis. She stopped eating and would have died had not my mother and I force-fed her barley gruel. In spite of her cat allergy, Mother would hold Petra in a Turkish towel to keep from getting scratched while I squirted the warm liquid down her throat with a medicine dropper.

Petra pulled through, and soon I was well enough to move back to my walk-up apartment in Greenwich Village. The only problem was, Petra had still not eaten solid food, and if she was going to have to be taken to the vet, I was not sure I could carry her up and down the three flights of stairs. I put dry food in her bowl every day, but she left it there untouched.

My parents and I were discussing this problem over lunch as Petra lay beside my chair. My mother didn't feel she could take care of the cat in my absence.

My father said, "Well, there are places where you can board animals. Maybe we should look into that."

As soon as these words were out of his mouth, Petra got up and walked into the kitchen. Soon we heard the crunch, crunch of her teeth chewing the kibbles that had been left in her bowl. She did not propose to be boarded in any kennel! My father, the scientist, may well have regarded this as an uncanny coincidence. I was inclined to believe the cat had understood.

Although this is Daisy's story and she never cared much for him, I really can't leave Max out of it.

Max was a white German shepherd whom his owner, a local printer, was planning to euthanize. Max was a special friend of mine, so I adopted him and cared for him for what turned out to be the last six months of his life. Daisy was friendly toward him at first, as long as she thought he was just visiting. But when on the morning after his arrival she perceived that he was being given a dish of food alongside of hers, she withdrew her welcome and would have nothing further to do with him.

Poor Max! I soon learned that he had that hip dysplasia that afflicts his breed. In spite of his pain he would somehow

get up the two steep steps to the cabin just so he could spend the night with me. Daisy slept in the house where she could keep an eye on things. But even though she didn't care to sleep in the cabin, she resented his presence.

I thought I detected a slight melting, a hint at the possibility of *détente*, toward the end. She no longer bared her teeth and growled at him in the kitchen; he had learned never to go near her bowl. And sometimes I even caught her sniffing his rear end, as if they had just met, in what seemed a sort of canine version of establishing diplomatic relations. I think she would have accepted him if he had lived longer. And Max, noble creature that he was, would have befriended her at a moment's notice, no questions asked.

I was spared the sight of his pitiful last days. My brother Warren and his wife Marianne had agreed to take care of Mother over Christmas that year, so I had been able to go to Woodstock. A few days after Christmas I slipped and fell on glare ice in a parking lot and fractured my pelvis. From my bed in Kingston Hospital I learned that Max was in very bad shape.

Sharon, who sometimes stayed with my mother at night, had taken him to the hospital. She said she had found him in the morning standing next to his doghouse, leaning against it for support. The dog knew that if he lay down, he could not get up again. The vet said there was nothing more he could do for Max. With a heavy heart I asked Sharon to call the vet and arrange to have the dog put to sleep. I felt bad that I couldn't be with him when it happened.

Sharon had a dog of her own. She offered to go there and be with Max in my place. The vet said he would put the dog down at noon. I called Sharon to tell her what time to be at the vet's. During our conversation I distinctly heard a dog bark—just one short bark. I assumed it was Sharon's dog.

"Was that Muttley?"

"Was what Muttley?"

"Didn't you hear a dog bark?"

"Muttley is asleep. I didn't hear a dog bark."

For some reason I looked at my watch. It was ten thirty.

Later that morning the vet called to tell me he had euthanized the dog. It turned out he had jumped the gun. He had forgotten to wait for Sharon. He said he was sorry, it had slipped his mind. I asked him what time he had given Max the shot. He said, "Ten thirty."

Some would call this coincidence, but I knew that single, clear bark I heard over the telephone was Max telling me goodbye.

In June of that year Mother fell on the tile floor of her bathroom and broke her hip. She was ninety-seven years old. Unlike most people her age she survived surgery, came home from the hospital, and lived three more years. Her stamina, intelligence, and longevity were an inspiration to her family and friends. My standard joke was that Mary Weaver was a product of the nineteenth century and that since then they had cheapened the model.

But Mother had lost too much muscle tone, and she never walked again. Gradually the bedside commode gave way to bedpans, urinary catheters, baby food, diapers: the long, slow descent into second infancy. As the only one in the house with limited olfactory sense, the result of chronic allergic rhinitis, I was often called upon to bathe "The Royal Bottom," as we called it, a job I found oddly satisfying.

If you think that humor doesn't belong in the same room with the dying, that it can't coexist with love and compassion, think again. It's how everybody gets through it, including the patient. Now that Dolly was no longer with us, the responsibility for comic relief had fallen to me and, of course, to Daisy. As Mother's care became more difficult the dog did her share by keeping us entertained with her antics.

As a young dog Daisy was healthy, and I seldom had to take her to the vet. There was the time Irma gave her a pork chop bone, and it got stuck in her throat. Daisy was so frightened she insisted on sitting on my lap in the car, thirty-five pounds of hysterical dog wedged between me and the steering wheel, the longest twenty minutes of my life. Of course the vet just reached in and pulled out the bone, no big deal.

Sometimes in the summer she'd chew on a fleabite until she had a hot spot. Then she'd have to wear one of those so-called Elizabethan collars. On Daisy the effect was more like the nun's headdress in Fellini's *Amarcord* and made her look as if she had taken holy orders.

Irma overfed the dog, and the vet warned that this increased the likelihood of leg problems down the road. About once a week, Irma would make a Perdue Oven Stuffer Roaster. After lunch she would put the roasting pan down on the kitchen floor and let Daisy lick up all the drippings in the bottom of the pan. When I glared at her, Irma would stick out her chin and inform me, "A dog's mouth is *clean*," as if she were quoting scripture. But that wasn't what bothered me.

After Irma left, I started feeding Daisy myself, and that was when she really became my dog. Gradually she resumed her pre-Irma weight. My brother and his wife, on one of their visits, pronounced her "positively svelte."

As she and I both aged, bathing Daisy got to be too difficult for me, and I'd take her to the vet for grooming. She hated riding in the car. She would shake, howl, drool, fart, and shed all the way down Route 7. She was always a little calmer on the way back home. And when she had to be in a kennel for the weekend, she barked so much that she'd be hoarse when I came to pick her up on Monday. Later I found a wonderful woman named Flora who boarded Daisy in her home along with her Lab puppy, Chloe.

In her early teens Daisy began to show signs of senile dementia, or what my friend Miriam called Dogheimer's. Instead of asking to be let out a few times a day, it would be in again, out again Finnegan, until it drove us all nuts. After Mother died, her behavior became even more erratic.

The Daisy Sutra

When I took on the job of caring for my mother, I told myself I would try it for three years. It never occurred to me that she would live so long. I also never dreamed that Mother and I, who did not start out as great friends, would become just that.

I watched in amazement as this woman who had been undemonstrative, anxious, and controlling all her life opened up in her late nineties to become affectionate, spontaneous, and trusting. Part of this transformation could certainly be attributed to dementia, but so what? For the first time in my life, my mother was showing me the child she had been, who was not that different from me.

It was close to the end of her life that Mother and I had our epiphany.

More than once, she had confided to the aides how disappointed she was that I had never had children, and of course they passed this on to me. On the day Mother made the mistake of saying it to my face, I exploded.

"You'd better be glad I didn't have children. If I had children, I wouldn't be here!" I screamed at this old lady in her hospital bed. "I never wanted children. I had something else I wanted to do."

"Oh, I know, you have compensations."

"I don't have compensations! I have the life I chose!" I shouted. I almost said, *You weren't my role model, Mother. Daddy was.*

Instead I said, "When you tell me you're disappointed in

me, how do you think that makes me feel?" I was close to tears.

"Oh, no. I'm not disappointed *in* you. I'm proud of you. I'm disappointed *for* you, because having you and your brother was the best thing that ever happened to me. I just wish you could have had that experience."

"But Mother, that's not what I wanted. I knew there was something else I was supposed to do. And I'm doing it. I'm writing. I couldn't have done both. I made the right choice."

Mother was sitting propped up in her hospital bed. I sat on the side of the bed, a valley of pink blanket between us. As we sat there across from each other, something happened that for lack of a better word I can only call a vision.

I saw her standing on the top of a mountain waving a blanket—maybe the same one that lay between us—over a fire. I was standing on a mountain of my own, studying her signal. We stood there, each on our own mountaintop, between us the gulf that had been there for sixty-some years, each of us sending smoke signals in the desperate hope that although we were not of the same tribe and did not speak the same language, we still might be understood. It seemed that for the first time in our lives we had each other's attention, and because of that, each of us was listening.

I looked at her across the gulf and for the first time in my life I thought, *Oh! So that's what it feels like to be you.* And I could tell that my mother was thinking the same thing. *So there's another way to be human, another way to be a woman, and that's OK. Your way and my way are both OK.* Maybe what I had been hearing all those years had not been a judgment after all. At least it wasn't a judgment any more. *Not disappointed* in *me, but disappointed* for *me. Oh.*

Friends at last. All because I screamed at her on her deathbed. I was glad I had. We read each other's signals in time.

🐾 One day not long after that, Mother asked the aide on duty to call me in. "She says she wants to tell you goodbye."

I went in and sat on her bed as usual. Mother looked me in the eye and told me she would be leaving soon.

I took the cool white hand that lay on the pink blanket in both of my own and returned her gaze. At age one hundred, my mother's unwrinkled face was even more beautiful than it had been when she was a young woman.

I knew she needed my permission to go. I said, "I'll miss you, but I'll be fine."

She said, "I'll miss you, too, but I'm not worried."

I almost fell off the bed. In the sixty-four years I had known my mother, I had never once heard her say she was not worried. Worry was her middle name, it was her job.

"Mother, what an accomplishment!" We laughed and hugged.

A few days later, on a Friday evening in early May, her favorite time of year, Mary Weaver took her last breath. The dogwood, the lilacs, and the star magnolia were all in bloom, laying a balm on my heart and helping me to face the whirlwind of change created by her sudden absence. Mother, with infinite tact, had chosen the perfect time for her departure.

I had told Mother I'd be fine, and basically I was. I had had fifteen years to prepare for this moment, and I thought I'd done most of my grieving ahead of time as I watched parts of her personality slowly disappear. The week after she died I lost nine pounds, too busy talking on the phone to eat, too wired to sleep. I put my all into her service, and it was exquisite, with Bach sung in German and Mozart in Latin, the two languages besides English that my mother knew and loved.

Then there was the house she and my father had designed along with the architect, their dream house, to empty and sell. My four nieces and their mother came from Kansas, Pennsylvania, Virginia, and Vermont and we had a warm, chaotic few days dividing up the linens and china, the books and art. They brought their children, and Daisy was happy to play with them and to gobble up their spilled food and leftovers. But she knew something strange was going on, and beneath her excitement there was an undercurrent of anxiety.

I was anxious, too. All those years I had assumed I would move back to Woodstock when Mother died. But now that she was really gone I was not sure where I wanted to live. The ocean called to me, and I spent a couple of days looking at houses on the coast of Rhode Island. But after a particularly depressing day I suddenly realized, *This is even further from my friends in Woodstock than New Milford was. What am I doing here?* My first love had been the ocean, but the great hearts of Woodstock had made me a mountain girl. At last, I knew where I was going.

Mother's death was as traumatic for Daisy as it was for me, for different reasons. I preferred living alone, but Daisy was gregarious. She missed the constant coming and going of nurses, doctors, family, and friends. I think she feared change, wondering what it meant for her. If it was true that she had been abandoned after her previous owner had died, she may have feared that history would repeat itself.

When the house was sold, and I began the enormous job of packing the possessions of my packrat parents as well as my own—over two hundred and fifty boxes—Daisy watched these proceedings with a glazed look in her eye. It was during this time of transition that I took the picture on the cover of this book. I can read the concern in that face that is looking right into the lens of the camera as if to say, *What's going to happen to me?*

I tried to reassure her. I promised her that she was coming with me, that I would never abandon her, that she was part of the plan. I told her we were going to live in the Catskills, which were real mountains, not like the Berkshires. In Woodstock the dogs wore bandanas around their necks. I told her I'd get her a red bandana, and she'd wear it around her neck, and she'd be a Catskills dog.

And I'd be a Catskills woman again. Now that I was leaving, I realized that Connecticut had been good to me. It had given me dear friends, new skills, and an unlooked-for intimacy with my mother. It had given me Daisy.

I had stopped calling Mother's town "New Mildew." I had stopped thinking of her house as "my beautiful prison." On moving day, seeing those echoing rooms empty for the first time in the four decades since their walls had been built to my parents' specifications, their very dimensions an expression of their excitement, love, and hope, was another death. *Partir, c'est mourir*, my mother's friend Mollie used to say. To say goodbye is to die.

Nobody in the family could afford to live there. The dream house had to be sold. I tried not to think of it as a betrayal.

The buyers, Roy and Judy, made several long visits before the closing, bringing their two daughters, their steel rules, and their excitement. Daisy had to be where the action was, and she was constantly asking to be let in or out the front door. It got to be annoying.

"She goes with the house," I quipped, not meaning it for a second.

"That could be a deal breaker," Judy shot back.

Moving in January was bad enough—she howled all the way, two hours over back roads—but Daisy and I had to do it twice. I couldn't face two real estate closings back to back, so

I rented for my first six months in New York. I wanted to give myself time to look for a house to buy.

My brother, whose health had been failing for several years, waited until I had sold Mother's house and was safely back in Woodstock, and then he died, too. My nieces were scattered in four different states. In a very real sense, in the sense of daily intimacy, Daisy was my whole family now, my only living link with the past.

That first winter in Ruby, a tiny hamlet outside of Kingston, was hard on us both. Daisy was now fifteen, and her eyesight was fading along with her hearing. The rental house, situated on ten acres of mature woodland, was pretty and peaceful, but Daisy and I were both disoriented and tired.

I was grateful when the one other tenant in the house, Elaine, turned out to be an instant friend. Tall and slim with a boyish haircut and a natural elegance, Elaine was one of those people who do everything not only well, but fast. I hired her to help me unpack, and she was a whiz.

Of course, she fell in love with Daisy on sight, and the feeling was mutual.

One day shortly after we moved, I made the mistake of letting Daisy out on her own. Elaine was helping me organize the kitchen. She was sure the dog would stay close to home. But Daisy wandered off. Three hours later she still had not appeared.

We called and called. We drove around Ruby. We even found some kids who had seen a little black dog, but we didn't find Daisy. Neither of us slept much that night. It was forty degrees out and raining. We both prayed that she would find shelter.

The next morning on her way to work—by some instinct, she took a different route—Elaine saw Daisy by the side of the road. She had to lift her, wet and trembling, into the back seat of her car.

Elaine came to my door, smiling, with Daisy in her arms. "Look what I found!" I was never so glad to see anyone in my life.

We dried Daisy off and I made a fire in the wood stove and set her up in front of it. We soon discovered that she was unable to get to her feet. Once up, she could walk stiffly, but we had to lift her with a towel under her belly.

At Kingston Animal Hospital the vet gave her something for arthritis, and gradually she made a good recovery. Her old friend Bill wasn't surprised. When I called him and Zada he said, "She's a compact, rugged dog." But Daisy could no longer jump into the back seat of my car, and I couldn't lift her by myself, so we went into a new phase. And of course there was no question of letting her out loose. From now on she had to be on leash.

A pet's life is so much shorter than our own. We know this when we take them on. A pet is a lesson in letting go, a home course in Buddhism. To have a pet is to embrace impermanence and to say, Yes, I will lose her. She won't live forever. But I'll do it anyway, because of the love. The love between an animal and a human is like no other love in the world.

Woodstock

By the time we got to Woodstock Daisy was an old dog. It wasn't easy finding a house for us. At first I had told Natalie, my agent, that I wanted an old house with charm. But after looking at some scary crawl spaces and moldy basements that would have been hell on my allergies—one had an actual stream running through it—I changed my tune. I started asking for something high and dry. By mid-April I'd been looking for three months. The lease on my rental was due to run out in June and I knew how long it takes to get to a closing. I knew my house was out there waiting for me, but I wasn't finding it.

I decided to pray to my mother. I said, "Mother, I really need to move this spring. Please help me find my house." An inner voice told me to write down exactly what I wanted. So I did: Seven minutes from Woodstock, 1800 square feet, two acres, fireplace, and so on.

Two days later Natalie took me to a house she'd had in mind for me all along, but for my phobia about ranch houses. It's a two-story ranch, but pretty, and as we walked through it it started reminding me more and more of my parents' house in New Milford. When Natalie opened the door to the linen closet I felt their presence so strongly my eyes started to fill. It was as if I could feel them bending over me and hear my father saying, "Now, this is more like it!" and my mother agreeing, "This is the house for you." Then Natalie opened the door of the master bedroom, and I recognized my office.

I had always sworn I would never buy a ranch house. Never say never.

The house was exactly seven minutes from Woodstock. It had everything else on my list, only instead of one fireplace, it had two.

This is a good example of what my friend Carrie calls "oogley boogley," a very useful term for those events you can't explain rationally, those so-called coincidences or sudden flashes of insight that give you goose bumps. Things were to get a lot more oogley boogley after I moved in.

The timing on my rental was iffy and I was nervous, but Walter at Allways Moving told me to relax and Trust the

Universe. That was when I knew I had finally come home. When the moving man tells you to trust the universe, you *know* you're in Woodstock.

Of course I took Daisy to our new home first to make sure she could go up and down the stairs. She passed the test, but after we moved in, the arthritis in her legs and back got worse. It got harder and harder for her to climb the stairs.

The medication for her arthritis upset her stomach, and sometimes she was incontinent. In December I thought I was going to have to put my dog down for Christmas.

Then Harriet, who had two dogs of her own, said, "Have you tried acupuncture?"

I called Howard Rothstein of Saugerties Animal Hospital, the acupuncture vet. Daisy started having weekly treatments, and they helped. Acupuncture gave her a new lease on life.

I almost wrote "a new leash on life." The leash that I had only used for occasional long walks in Connecticut was now a necessity, part of our daily routine. I didn't feel I could afford to put up a fence around the front yard, so Daisy could no longer run free. The up side of this was a new level of intimacy between my dog and me. The fancy retractable leash Marianne had given me for Christmas wasn't just a practical solution; it was an umbilical cord that bound us together as never before.

My day began much earlier now. And I, who could spend hours glued to the computer or with my nose in a book, unaware of my body or the weather, had to be out walking at regular intervals throughout the day. Daisy's enthusiasm for nature, for the changes of the seasons, the warmth of the sun, the smell of the grass, the magic of snow, was infectious. The way she sniffed at the ground, following some animal's scent, getting the news of the day, seemed so much healthier than humans' secondhand reports from the media.

If it was a sunny day, I would sit on my favorite stump and sun my eyes while Daisy sat—not waiting, because there was nothing in the world she would rather do than sit beside me in the sun, but just being alive in the moment. It was our way of meditating, a kind of communion.

If it was raining, I would get out her red vinyl dog raincoat, slip the hood over her head and secure it under her belly, and off we'd go. It was a little too big for her and the effect was quite fetching.

Daisy had a characteristic way of lifting her front legs when she walked. I don't know if it was her beagle blood—the lifted paw suggested a hunting dog at point—or simply an expression of her own ladylike personality, but I found it utterly charming.

Every once in a while she would stop in her tracks and stare into space as if she saw something that was invisible to me. She would just hang there for no apparent reason until I urged her to move on.

My Buddhist friend Michael and I were walking Daisy one night by flashlight when she went into her canine trance.

Woodstock

"What is Daisy doing?" Michael asked.

"Absolutely nothing."

"You mean she's just staring into the Void?"

"That's exactly what she's doing."

We had our routines and our favorite spots, but every day was different. And of course there were days I really didn't want to get out of bed or interrupt my work. But once outside with Daisy my mood would change. I would inhale the air and look around at my two acres in Woodstock and know that I was blessed, that I was the luckiest person in the world.

And I'd sing to her as we walked, and call her all those silly names. And of course I'd praise her to the skies when she accomplished the practical, as opposed to the spiritual, purpose of our walk.

Because she did sometimes make mistakes in the house. Her digestion was not what it used to be. There were even times when the great eater turned up her nose at dog food.

One evening, to perk up her appetite, I put some chicken fat in with her kibbles, lamb and rice. That night she woke me up at 3 A.M. asking to go out, but it was too late. She had messed all up and down the hall. I spent the next hour on my hands and knees cleaning up fifty-three little explosions on the carpet. I counted them to keep up my morale. Daisy had lost the ability to process fat, a sure sign her body was starting to break down.

The Buddha's path to awakening began when, as a sheltered young prince, he first left the palace and saw a sick person, an old person, and a corpse. His search for enlightenment was prompted by his discovery that illness, old age, and death come to all sentient beings.

The stages of Daisy's decline mirrored my mother's. The vet said to give her baby food. Finding myself in that aisle of the supermarket again was a painful reminder of the last months of my mother's life. Once again I was caring for an elderly female.

Daisy became a favorite with the staff at Saugerties Animal Hospital. Of the four vets I had taken her to—two in Connecticut and two in New York—this was the one where she felt most at home. At both of the places in New Milford she had been too nervous to pay much attention to the other animals. Here her regular visit was a social occasion. She always checked out everyone in the room along with their pets. When she went over to someone, I always told them proudly, "Her name is Daisy, and she's sixteen." They were always amazed at how well she looked.

One lady said, "She's the perfect size for a dog—not too big, not too small." I had to agree.

The arthritis in her spine made it hard for her to get to her feet at times, and in the spring I got her a harness so I could help her up and support her on the stairs. Sometimes she would whine softly with each upward step. Feeling guilty, I told her she was a champ.

Once too often she fell down the carpeted stairs to the basement before I could reach her. I would watch, hurting for her, unable to help. She always picked herself up and was ready for our walk anyway, though it couldn't have done her arthritis any good.

Her bravery amazed me. Once outside, arthritis or no arthritis, she would run. She had to do her running on leash because I didn't have a fenced-in yard, but we had a system. I would stand in the middle of the front yard and pivot while she ran around me in a wide circle, not looking like an old dog at all.

Sam, my new handyman, sometimes house-and-dog-sat for me, and he became Daisy's buddy. Under his many tattoos and body piercings Sam is a regular guy. Tall and with the kind of rugged good looks sometimes called Lincolnesque, Sam always took time to squat down and pet Daisy. He was one of the few people for whom she still wagged her tail.

Sam said she reminded him of the great football player Jim Brown. He said Jim Brown would come out of the huddle looking like he was on his last legs, but once he had the ball, he would streak down the field like a bullet.

Inevitably the day came when Daisy refused to climb the stairs. I knew it hurt her to do it; now it hurt too much. I knew what I had to do.

I brought her food and water bowls and her fake sheepskins down to the first floor. I don't want to call it the basement. It's bad enough my dog had to live the last months of her life on the ground floor, without calling it the basement. It isn't damp, it's carpeted and completely furnished, it has windows and a wood stove and a couch in front of the TV. But it doesn't have my kitchen, my bedroom, or my office. For too many hours of Daisy's day, it didn't have me.

Daisy's new room—I call it the den—was off the garage, so now I could walk her more often without having to worry about stairs. But it wasn't the same, and she missed me. She lived for the times when we were together: her two meals, our five or six walks, and the evenings, when I would bring my supper down to the den and watch TV. If it was five minutes to seven and I hadn't appeared, Daisy would let me know it was time for The X-Files.

In between those times she was lonely, or maybe just bored, and all too often, she would cry. "Inappropriate vocalization" the vet called it, a sign of senility. Mother used to talk to herself for hours. I don't know how inappropriate it was, but it broke my heart. It made me feel guilty, and at times it drove me crazy. I would yell down the stairs for Daisy to be quiet. Usually she stopped crying at once, as if reassured that I was home.

But sometimes she wouldn't stop. Once, in exasperation, I threw a slipper downstairs, not at her, but just to get her attention. Later when I went down to take her for a walk, Daisy was lying with her nose on the slipper, fast asleep. The object I had thrown in annoyance had comforted her, because it smelled of me.

Daisy had a low-grade chronic colitis, for which the vet had prescribed an antibiotic and a diet of white rice and chicken. She wolfed down her food again, as of old, but she had lost a pound or two and was sometimes incontinent. Even with the acupuncture treatments, she moved stiffly.

She wasn't the dog she used to be; but was she really ready to die?

I had heard that there were people called animal communicators who were somehow able to tune in to the feelings of nonhuman animals and translate those feelings into words a human could understand. Two friends in my women's group had experienced this type of communication with their pets.

Shellie's dog Peaches, a pocket-sized Cairn terrier—the same breed as Toto in *The Wizard of Oz*—was in distress.

Shellie called animal communicator Gail De Sciose in New York City and made an appointment. Gail can contact an animal while speaking with the animal's person over the telephone. Gail said Peaches told her, "My right ear itches." Shellie inspected both ears and found a nasty fungus in the dog's right ear. The left ear was fine.

Animal communicators believe that animals, like humans, have a spiritual essence that survives death. For communication to take place, not only does the animal not need to be present; she need not even be living. Through Gail another friend, Karen, had several conversations with her dusty calico cat Perdita both before and after Perdita died. Karen was so moved by the experience that she decided to take a workshop in animal communication.

I called Gail De Sciose. I wanted to know how Daisy was feeling, whether she was ready to die, and if so, whether she needed help or wanted to do it on her own. Gail suggested I prepare a list of questions for Daisy and things I wanted to say to her. We made an appointment to talk on the phone. When the time came, I took my list down to the den.

FIRST CONVERSATION WITH DAISY, JUNE 11, 1998

When I called her, Gail asked how Daisy was and what she looked like. After I had described her, Gail took a moment to center herself and tune in to Daisy, who was right beside me. Gail said that often animals become quite relaxed during a session, but not to be surprised if at first she looked around, trying to see Gail.

This is exactly what happened. At first Daisy looked all around as if wondering where this person was. Then she settled down on her sheepskin.

After her meditation Gail said, "She's so cute! I explained to Daisy that I am talking on the telephone to you, and asked if she understood. Daisy says, "*I understand the telephone. I have seen quite a few. I can talk to you.*"

"Please tell Daisy that I love her very much."

Daisy says she loves you, too. I get a strong sense of curiosity from her."

"How does Daisy feel? Is she depressed?"

"She says, *Everything in my life goes more slowly now. I move more slowly, time goes more slowly. I am happiest when Helen is here with me.*"

"Is Daisy ready to die?"

"She says, *I am not ready to leave yet.*"

This did not surprise me at all. "When she is ready to go, will she want help, or will she prefer to do it on her own?"

"Daisy says, *I think I can do this on my own, but if I'm having trouble leaving, then it will be okay for you to help me.*"

Please tell her that no one is hurrying her along."

"She says, *Thank Heaven for that.*"

"Please tell her that if and when she does need help, she will be in her own home, not in the vet's office."

"Daisy says, *That would be good.*"

"How does she feel about the acupuncture treatments?"

There was a pause. "She says that at first it was difficult holding still with the needles in, but now she likes it. The person who holds her during the treatment pets her and talks to her, and she loves the attention.

"Daisy says she loves to go outside in the grass and in the sun. She is showing me a picture of you reading and her sitting on a blanket next to you."

Concerned over her recent weight loss, I asked if she was hungry.

"Daisy says, *Sometimes I'm hungry right after I eat.*"

This was an eye-opener for me. After that I doubled up on the amount of chicken, increased the rice, and, at the vet's advice, added carrots and broccoli to her two daily feedings.

I said, "My father used to say he could never understand the expression 'It's a dog's life.' He thought that a dog in a nice family had a very good life. Daisy was born a few years after Dad died. I've often wondered if Dad reincarnated as a dog in order to keep an eye on Mother and me."

"Daisy says, *I am not this being.*"

"Please tell her that she is not in the basement as a punishment. She is in the basement because she can't climb the stairs."

"She is showing me a picture of you lying on a couch in the basement and her lying beside you."

Gail had no way of knowing that my basement had a couch in it. Daisy told her that!

"Please tell Daisy that I am very proud of her. She is a very special dog. I want her to know that I am writing the story of her life. She was a great hunter and she was also a good hunter, because she ate every animal that she caught."

"Daisy thanks you. She is very pleased about the story. As for being a good hunter, she says, *Yes, I was, but I'm too slow to do that now.*"

"I feel bad because I know she misses me when I'm upstairs."

"Daisy says you can move to the basement. I told her that even if that were possible, you would still have to go out to work, buy food, and so on. She wishes you could live in the basement. But like all animals, she lives in the present and accepts what is."

Gail recommended giving Daisy the Bach Flower Remedies, homeopathically prepared essences used for emotional healing. I was already putting Rescue Remedy in her water. Gail said that a few drops of Honeysuckle and Walnut Flower Essences in her water would soothe her when I had to leave her alone or with Sam.

That was June. Daisy wasn't ready to die, and I wasn't ready to lose her, but I knew in my heart it wouldn't be much longer.

Woodstock

When Annie had visited us early that year, Daisy no longer recognized her. There was no doubt about it, because back in Connecticut, whenever Annie came for a visit, Daisy would carry on as if possessed, moaning and jumping in a frenzied and protracted ovation that both touched and tried the patience of her special favorite, who really preferred cats.

That summer I had a visit from my Connecticut friends Bill and Zada. By now they were both retired and had sold their house in Brookfield and were traveling the country in their mobile home. Bill and Daisy had always been an item, so I told him not to take it personally if she didn't know him.

At first it looked as if she didn't. But Bill made a big fuss over her, and after he had been petting her for a while, something—maybe his scent—broke through the fog of her faded senses. She began to moan ecstatically, a sound so expressive of either pleasure or pain that I couldn't tell if it was joy in the presence of her old friend or grief for her lost youth in Connecticut which that presence had evoked.

Early in July my friend David visited me from the city with his new guide dog, Siri. David is legally blind and suffers from occasional panic attacks. Before he got Siri he rarely traveled. Siri is a beautiful long-coated German shepherd, black with tan markings, and although Siri was two or three times her size, Daisy accepted her at once. Daisy seemed to understand that Siri was David's dog and therefore not a threat. And Siri, for her part, seemed to sense that Daisy was elderly, and treated her with great sensitivity and respect.

The two dogs enjoyed our walks together under sun, moon, and stars with the crickets booming around us. David and Siri spent the night in the guestroom on the first floor, right off the den where Daisy slept. When I came down to walk her in the morning, I found her curled up right outside their door—the ultimate seal of approval.

Daisy had always been afraid of thunder—maybe more than most dogs because of the times she had spent exposed to the elements. During an electric storm she would come to me terrified. I would hold her close and tell her I wouldn't let anything bad happen to her. Now she was too deaf to hear real thunder but she could still hear the fake thunder on my

CD of *The Magic Flute*. She quivered and quaked along with Papageno in the second act.

Now when I stroked her body, I could feel her spine and her rib cage. "They waste," the vet had said. I didn't believe it would happen to my Daisy, but it did.

I could still make her groan with pleasure by massaging her ears a certain way, but it had been a long time since she had really wagged her tail.

After Mother had stopped speaking to us, had metaphorically turned her face to the wall, I could still call forth once again that smile I so badly needed by bringing her roses. It's called denial.

Daisy's back legs were now so weak that they no longer supported her when she had serious business to accomplish. By pulling up on her harness, I helped her maintain her dignity.

She still ate well, but she had stopped drinking water. Elaine, who came to clean my house every two weeks, and Sam, who also saw her regularly, knew that my dog no longer had that elusive thing called "quality of life." I, of course, was the last to know.

Then one morning in August, Daisy told me herself. When I went downstairs to walk her I found that she had soiled her bed, and she couldn't get up. A voice in my head said, *It's time.*

Oh God, it was so hard. I had never put an animal down before. She was my best friend. She had stuck by me all this time, had kept me company all those long years in Connecticut, had put up with two traumatic moves, had seen me safely back home, had given my life continuity, had agreed to do all this in spite of her pain. How could I let her go?

That morning, a Thursday, I called the vet's office. Howard Rothstein said he was willing to come to my house to help Daisy die, but he would need a few days' notice. It turned out he would not be able to come until Monday.

I was so afraid I would change my mind and we'd have to go through this all over again that for a moment I pictured taking Daisy to the vet's office in the car like most people do—coming with a dog and leaving without a dog. But that drive was so stressful for both of us, and I had promised her she would die at home.

I felt like I was doing this for me, because her care had become so difficult. I was so afraid Daisy wasn't ready to go that I didn't dare call the animal communicator. I didn't want to know what Daisy was feeling. I felt so guilty. I was a mess.

I called my boss and told him I'd be late for work because I was upset about putting my dog down. He said cheerfully, "Okay, I'll see you when you get here." Shocked at what seemed like callousness, I confronted him. He said he hadn't understood me. He had never heard the expression; it must be regional.

I had heard the expression often, but I had never had to do it. When Brownie, the dog of my childhood, had to be put to sleep, I was in college. It must have been my stoical mother who had to make that last drive to the vet.

My two cats, Misty and Petra, both went out of my life while still young. My subtenant, not a cat person, gave Misty away when I overextended my stay in Europe and the cat began messing the bed in protest. As for Petra, whom I brought home with me from Greece, and who saw me through one of the worst times of my life, I had to find a home for her after I developed an allergy to cats. That was the hardest thing I had ever done in my life—until now.

I had been spared the sight of Max's agony and the responsibility of his death. This was a first for me, and I needed all the help I could get.

I called Elaine and Sam, Daisy's best friends in Woodstock, told them what was happening, and asked if they would be willing to come and be with Daisy and me when the vet came to our house. Somewhat to my surprise they both said, "You're doing the right thing. She's had a good life, but it's time." They both promised to come.

If I was able to let go for a few moments of the guilt about putting my dog to sleep I would immediately be consumed by guilt for not doing it sooner, for keeping her alive too long and letting her suffer. It was a no-win situation.

On Saturday morning I woke with a strong need to communicate with Daisy. I called Gail De Sciose. The message on her machine said she was away for ten days and gave the name and phone number of another animal communicator in New Jersey. Instead, I called my friend Karen Beth.

Karen is a singer, songwriter, and massage therapist. She is not a professional animal communicator, but she had taken a workshop with Gail and she has the gift. I told Karen I had promised Daisy that she was to be in charge of her death, and I did not want to betray that promise. I wanted to know how she was feeling and whether she needed help in leaving.

Karen works a little differently from Gail. Instead of acting as an interpreter, she prefers to contact the animal one on one. I gave her a list of questions over the phone. Karen was able to tune in to Daisy, and called me back with Daisy's answers. Later she sent me a written report. Here it is:

Second Conversation with Daisy, August 15, 1998

Karen: Hello, I am Karen, Helen's friend. Helen would like to know how you are doing.

Daisy: My body grows weak. My spirit is strong and can be set free.

Karen: Helen loves you so very much.

Daisy: As I love Helen. When you have a relationship such as Helen and I have had, there is no good time to say goodbye.

At this point [Karen wrote], I see her really

angry at having to die—baring her teeth, growling. I tell her what it was like for my cat Perdita before her death, and how she was so free and happy after leaving. After she died Perdita told me, "I can hear again! I can hear the trees growing, the little birds' hearts beating." Her joy was intense.

Daisy: Thank you for reminding me that there is no death, that I will be free and able to be with Helen from the other side. I have had a good life. I will allow help in my leaving, whether it be by angels or Helen's help.

I see a light tunnel near her going off into infinity.

Daisy: Please tell Helen how very much I love her. I've enjoyed our walks together, sitting near her, watching out the window, the different seasons, being in the snow, playing with leaves. I love the outdoors. It is hard to leave, yet I know I must. I have heard the call and didn't want to listen and now I know I must surrender.

It is hard to leave for I love life so. I will return to a body again, if possible. I will stay in touch with Helen. We will never be parted. She is my friend, my companion. I love her. Please tell her not to miss me too much, nor be sad for too long.

I see her in a doorway looking out at nature, at a field and distant tree, daytime, and stars in the sky at night. I see her running in a mowed field, sniffing at a big tree.

[Helen's note: There was a mowed field and two-hundred year old oaks on our property in Connecticut, which Karen has never seen.]

Karen: Daisy, you are a fine dog.

Daisy: I am and shall evermore be so. I can go now. I know I will be taken care of. Please tell Helen that I need to be held and loved. I need Helen to talk to me of our life together.

Thank you, Karen.

In spite of this conversation I still couldn't sleep at night. After two terrible nights of grief and guilt I woke up Sunday morning and said to myself, I can't do this. This feels wrong. It feels as if I am doing it for myself.

I called Karen to ask if she could tune in to Daisy again. She was leaving for work at Omega, a holistic learning center in Rhinebeck, New York. She suggested I call that New Jersey number Gail had left on her answering machine.

I called the number and, wonder of wonders, got a human being instead of a machine. Ginny Debbink listened to my cry for help and agreed to schedule a session with Daisy in an hour.

I fed and walked Daisy, had a bite to eat, and took the cordless phone and pen and paper down to the den so I could be with Daisy during the session.

THIRD CONVERSATION WITH DAISY, AUGUST 16, 1998

At Ginny's request I described Daisy and explained our situation: how I found her Thursday morning having messed her bed; how an inner voice told me, *It's time*; how I called the vet and made an appointment to have her put to sleep, and now felt I couldn't go through with it.

Ginny said she would take a few moments to center herself, contact Daisy, and ask her permission to speak with her. "The way this works is, if the animal has something to say, they have an opportunity to go first. If not, the human can start talking or asking questions."

"Please tell Daisy how much I love her."

"Daisy gives me her permission. She is surprised I am not Gail. She says she loves you too. Daisy has a lot to say. First, she thanks you for your patience. She says she is sorry for the mess. She knows she's not supposed to do it inside, but she couldn't help herself."

"It wasn't her fault!"

"Daisy says if she'd known how patient you are she might not have been such a good girl all these years! Daisy knows your questions, knows why you are calling, and wants to help you. She says for you to look into your heart."

I looked in my heart, and in it there was nothing but pain. I said, "I can't do it. I feel as if I'm doing it for me."

"Daisy wants you to understand something. You need to understand that you would never have picked up the phone and called the vet if she had not asked you to do it. She is pleased that you got her message. She is tired."

I told Ginny, "My friend Karen Beth talked with Daisy yesterday. She picked up some anger—not at me or at being put to sleep, but at death itself, at the fact that we all have to get old and die."

Ginny said, "Animals go through the same process as humans: stages of denial, anger, acceptance. Daisy has worked it through. I pick up no resistance in her. Daisy says she is tired of the humiliation. The anger she has felt over the past few months has been anger at the breakdown of her body. Anger at not being able to do the things she used to do. It is very important to Daisy that you know that this is an idea you got from her. She asked, and you heard. She sees your guilt. She wants you to clear the guilt. You heard her."

Ginny went on, "Every living body is tied to the Earth. Every spirit has memories of being out of the body, but we don't have those memories all the time, because we are here to experience life in the body.

Daisy is ready to leave. There is a longing for freedom, but we must all work through the hold of the body. Daisy says her body is tired. It's a weight."

I asked, "How can I make her last twenty-four hours in her body as good as possible for her?"

"Daisy says, *The chicken is good*! She says she wants toast with peanut butter. She says she wants time in the sun. She is sending me a picture of a pond."

Maybe that was a memory from her puppyhood. I asked, "Why doesn't she drink water?"

There was a long pause. Finally, "Daisy says, *It has no taste*!" (This sounded so much like my bourbon-drinking father I thought, *Are you sure you're not him*?)

"Some of the anger she has been feeling the past few days is because much of her body is still good, and it seems a shame. She is not in severe pain but she is worn down by low levels of constant pain. She is telling me again, this decision started with her. She knows that you would not be able to make this decision only for yourself. You have always given her what she needed."

"Please tell her I am proud of her, for so many reasons, but especially for her courage in bearing her pain without complaining."

"Daisy is grateful. She's glad you noticed. She

says that dignity is important to her."

"Please tell Daisy how much I love the elegant way she walks, the way she picks up her two front paws so daintily, like a highbred horse."

Ginny said, "This is called 'hackney gait,' lifting from the wrist. Daisy says it makes her very happy that you noticed. She is aware of her own elegance. She says, *I'm dainty and neat. I've always been tidy, even in the way I move.*"

And once again, as if my dog knew how much I needed to hear it: "She wants you to let go of guilt. It's time. She wants you to trust in your connection, that this message came from her. She wants you to receive her love and gratitude, and to know that she'll never be far from you."

After this phone call I was able to breathe again. I was able to eat a little something, to call some friends. I was so grateful to Ginny for her gift of being able to communicate with animals and to Daisy for her kindness and her love.

I got the traveling yoga mat I keep in the car and put it next to Daisy's bed in the den and lay on it with Daisy. I held her and petted her and told her stories about her life, recalling her hunting exploits, our walks together in the hills of home. I praised her heroism and thanked her for her loyalty and her love.

I called Harriet, whose terrier mix Chikita had died in her arms after a long illness. Harriet had a copy of a tape called *Animal Death: A Spiritual Journey* by Penelope Smith, an animal communicator who was Gail De Sciose's teacher. Harriet offered to come by that afternoon and bring me the tape and pay her last respects to Daisy.

I told Harriet about my conversation with Daisy, how she had asked to hear stories about her life.

Harriet said, "That's what the dying want. They want to remember their life."

Harriet had known Daisy almost as long as I had. We recalled the time back in Connecticut when Harriet had called Daisy's begging vulgar and Daisy had walked out of the room. Harriet was sure that Daisy had never forgiven her. I was sure she had.

I told Harriet about Daisy's request for toast with peanut butter, and Harriet offered to purchase a jar. Peanut butter happens to be my favorite food, too, and what I would want for my last meal. I like the organic kind from the health food store.

But Harriet said, "No, she wants Peter Pan." In the end, she couldn't get Peter Pan, so she settled for Skippy.

When Harriet arrived it was clear that, whether due to memory loss or to her innate forgiving nature, Daisy had put the past behind her and bore Harriet no ill will.

It was 95 degrees in my upstairs kitchen, so instead of toasting bread, I reached for the dog biscuits. We broke up the biscuits and scooped out small amounts of peanut butter with the pieces, and Daisy—well, she thought she had died and gone to heaven. Daisy had a tendency to grab a dog biscuit with her teeth but Harriet showed me how to hold the piece of biscuit in the palm of my hand "the way you feed a horse." After she'd had several biscuits we each put a dab of peanut butter on our palm and let Daisy lick it off.

Daisy had never been a licker. She confined her more intimate demonstrations of affection to sniffing your breath with deep concentration, as if making a precise classification of your personal bouquet. To feel her warm tongue enthusiastically cleaning off the palm of my hand felt wonderful.

We took her for a little walk outside. Daisy had asked to spend time in the sun but it was so hot and humid that Harriet and I were dripping and barely able to move. I put the portrait lens on my old Pentax and, while Harriet held the handle of her leash, took a few last pictures of my dog.

Then Harriet left, and Daisy and I were on our own. It was our last evening together, and every moment was precious. I brought my pillow downstairs and made a place to sleep for myself on the couch. I got the good yoga mat from the meditation room and dragged it down the stairs. I fixed it so Daisy could lie down right next to me if she chose and still be on foam rubber.

Daisy, like me, has always been rather private when it comes to sleeping. Back in Connecticut she had divided her night between her favorite spot in the foyer where she could keep an eye on the door and the club chair in the living room. (Theoretically, this chair was off limits, but Daisy's persistence wore me down and eventually I gave in and let her sleep there.) When the night cooled down toward morning, that chair was the perfect size and shape to hug her little body as she draped her chin across the arm rest.

After Mother died and I was alone in the house, I tried to get Daisy to sleep in my room, but she refused She would come bounding into my room in the morning and ask to be let out, wriggling with joy when I finally hauled myself out of bed and padded down the hall to the front door, looking over her shoulder periodically to make sure I was really coming.

Around six-thirty I walked Daisy as usual, fixed her supper, served it, made my own, and managed to get down some of it. For our last walk under the stars the crickets and katydids were playing their summer evening concert all around us and it had cooled off a little.

I asked her, "Can you hear them, Daisy? Isn't this a beautiful planet you've been visiting?"

The couch in the basement had seen better days, but I wasn't really expecting to sleep. I left the light on over Daisy's water bowl, as usual. This meant that I could see her every time I opened my eyes. I listened to the tape Harriet had brought. It was full of stories about animals Penelope Smith had contacted in her work as an animal communicator. She believes that animals have an awareness of their spiritual nature and do not fear death as most humans do:

> "With angelic beings like this many people feel so blessed—the animals are so wonderful and give so much love, peace, and light—that people may want to hold onto them. When beings like that come into your life, count your blessings. Appreciate them for who they are, and acknowledge the wisdom they bring. They may be here for a short or long time, according to their purpose. Honor them and let them come and go, knowing that you are always connected spiritually."

Woodstock

Monday morning dawned overcast. After another mostly sleepless night, I awoke with a knot of anxiety in my gut. I had made the appointment for Daisy to be put down with the vet's helpers, Anna and Lisa, without ever talking to the vet himself. Now in the paranoia of my exhaustion I imagined him coming to my house and saying, "She doesn't look so bad. Why did you stop the acupuncture treatments?" I knew in my heart he would say no such thing, but my tormented mind kept rehearsing this scene.

I called the vet's office. Lisa was very kind. She said, "He would never say that. Not in a million years."

I told Daisy I was sorry there was no sun on her last day. She enjoyed smelling the grass, and when I took her onto the front lawn, she even had a little run.

I remembered her request for toast with peanut butter, but stupidly, I didn't toast the bread. Of course the whole thing formed a paste and stuck to the roof of her mouth. Her struggle to get it down would have been comical if it hadn't been the day that it was. I quickly gave her half a puppy biscuit and that did the trick. *She* knew she needed something crisp. What dogs put up with from their so-called masters!

Daisy's appetite was good, and for the first time in weeks she actually drank some water. I wondered, was she was doing it to please me, or did the peanut butter make her thirsty—or was she saying goodbye to water? Whatever the reason, I was happy to see her lapping it up. At last she'd get the calming effect of the Rescue Remedy I'd been putting in it for weeks.

After her last walk I took off her harness and gave her a gentle brushing.

I had prepared a spot for her in front of the wood stove. The towel I put on the yoga mat would be used to wrap and carry her body later. I chose a beautiful red and lavender one that Miriam had given me one year for Christmas.

At last the terrible morning went by, and Elaine came and then Sam. I read them the report of my last conversation with Daisy through Ginny.

When the vet's car turned into my driveway, two people got out. Howard had brought a helper, Sally, and she was carrying a red muzzle.

Howard said, "I'm sorry. But it's time."

I said, "The acupuncture treatments helped her so much."

He said, "I'm glad we were able to help her."

I asked what the muzzle was for.

"In case she resists."

I said, "You won't be needing that. She's ready."

Inside I introduced everyone. Daisy got up from where she was lying and walked around the room one last time. I unfolded the red and lavender towel and spread it out on the mat. We called her over to the mat and she came.

Sally got behind the dog and put her arms around her body. Howard knelt in front of Daisy, got out his needle, and filled it from a little bottle in his bag. I had pictured myself holding Daisy, but now I got in front of her so I could look into her eyes and so that I would be the last thing she saw. She seemed very calm and relaxed. I kissed her velvet forehead, stroked her smooth nose, and told her goodbye. My hands were on either side of her dear, grizzled face.

Daisy gave a mild little cry as the needle went into the vein in her right front leg, very near the wrist from which she lifted in her dainty, elegant walk. She sank down, and the three of us helped her onto her side. She looked asleep in her most comfortable position, all four legs lying in the same direction out to her left. But when I embraced her body, my hands on her rib cage, there was no motion, no breath. One minute she was here with us, and just seconds later she was gone. I tried to close her eyes, but dogs don't have eyelids and her eyes remained half open.

Howard and Sally wrapped her tenderly in the colorful towel. Howard lifted her up and carried her out to the garage, where I had cleared a space for her on the back seat of my car. He had other calls to make, so I had to take her to Saugerties, to his office. Then they would send her to the crematorium in Hartsdale.

I planned to take some of her ashes back to New Milford and scatter them on the hills she had loved, the place she had been happiest. I was sure the new owners would not object.

I told Howard about the terrible mix-up when Max was put down in New Milford. That vet had not only jumped the gun, but had also lost Max's tags, so I could never be sure the ashes they gave me were really his. Howard said this place in Hartsdale will actually allow the owner to come there and watch the cremation process.

The drive to Saugerties Daisy and I had done so often together was dreamlike. At the vet's I opened the back door of my car and unfolded the towel and gave Daisy another hug and kiss.

A young boy carried her in the back door into a small room with cages all around, some of them with animals inside. In the middle of the room was a metal table with several dozen food bowls. I moved these to one end to make room for Daisy. The boy laid her down and carefully removed the towel from under her body. A cat jumped up on the table right next to Daisy and began inspecting her. The boy took the cat off the table, and I made him promise it wouldn't happen again. He asked if I'd like a few minutes alone with her. I said I would. I was having trouble letting go.

She looked so beautiful to me, and it wasn't just because I loved her so much; she *was* beautiful. Like my mother, she was elegant on her deathbed and elegant in death. I knew I had done the right thing and that she was at peace. But how could I bear never to see her again? Her eyes were already clouded over with a kind of opaque brown film. I kissed the velvet forehead again, felt the silky ears, the smooth nose, and told her goodbye.

Beyond

Tears come from a place somewhere at the back of your throat and rush upward to your eyes. There must be a muscle there that you don't use very often because when you cry a lot, that place aches. I kept thinking of Mother's phrase *lacrimae rerum*, the tears of things—was it Lucretius? Mother, the Latin teacher, would know. But she was gone, too. They were all gone now: my father, my mother, my brother, my dog. Now I was really alone.

At first, I would listen for the sound of her license and rabies tags clinking against her bowl and think I heard it. If there was a wheeze in my breathing, I would think it was her soft cry coming up the stairs from the basement telling me she missed me, or she was hungry, or it was time for a walk. I was happy that she was at peace and that I was relieved of the responsibility of caring for her and that that awful decision was behind me, but I missed her terribly. No contradiction.

I could cook for myself again now that I didn't have to cook for Daisy. I talked to her as I stood in the kitchen looking out the window. If the sun was shining, I'd tell her how sorry I was that the sun didn't shine on her last day. Then I thought, Maybe that made it easier for her to go, and I wept.

I remembered Daisy saying she hoped I wouldn't be too sad. I told her, "I'm all right, Daisy! Don't worry about me; but don't leave me. You promised you would never be far from me, and I'm holding you to that promise. Besides, I'll need your help with my book. *Our* book."

I thought of her down in the basement, alone and in the dark. "I know you have forgiven me for keeping you in the basement after you stopped being able to climb the stairs. I am trying to forgive myself. I do forgive myself. But I promise you this: If you ever decide to take a body again, and you come to me in *my* old age, I will try to give you everything you need and make sure you are happy. I promise that you, the sun lover, will never have to live in the dark and that you, the lover of grass and fresh air, can run around freely without having to be on a leash. I did the best I could, and you know that. You told me I gave you everything you needed. I don't think I did, but it was so kind of you to say it."

Beyond

A card came from Sam and Diane: "Not just a pet. A member of the family."

Friends called to comfort me. I told them about the conversation I had had with Daisy through Ginny Debbink the day before she died. I told them how her last request was for toast with peanut butter. I had lived with this dog for fifteen years, and never knew she loved peanut butter! Clarisse cried for Daisy as if she were her own dog.

I reminded Bill that he had once told me that Daisy was his favorite living dog. I had always assumed that there was some favorite nonliving dog whose memory he was honoring by making sure to include the word living. So I asked him if Daisy was now his *next* to favorite non-living dog. But Bill said that this was not the case. He said that Daisy was his favorite eternal dog.

The day Daisy died a cricket appeared in my bedroom. For three weeks she sang to me every night in the empty house, as if she knew I needed company.

The Daisy Sutra

Daisy's ashes came in a pretty metal canister, black with a pattern of red and white roses—perfect for a black and white dog who looked her best in red. I put it on the altar in the meditation room with her old red collar wrapped around its base.

Three days after Daisy died I came down with a terrible cough that was eventually diagnosed as acute asthmatic bronchitis and lasted for two months. It was the oddest illness I'd ever had. There was no cold, no runny nose, no sore throat—just this awful, convulsive, rib-rattling cough that went on and on until I thought I would have it for the rest of my life.

The oddest thing about it was, the place the cough came from was that same place in the back of the throat that tears come from. In retrospect my illness seems like a way of making space for grief. In the end, the cough was very much like a bark.

Eventually, I got better. I still missed Daisy, and I still talked to her and sang her the "Itsy Bitsy Baby Doggie" song. At the same time I was glad I wouldn't have to haul myself out of bed that winter and walk her on the snow and ice.

Sam asked if I was going to get another dog. At first I thought, No way. I'm not going through this again in my eighties. But I found—I still find—that every dog I see, in a magazine, on TV, or in life, goes straight to my heart, and out of my mouth comes an involuntary "Awwwww . . ." And I realize there's nothing like a dog.

"Does a dog have Buddha nature?" is a standard Zen

Buddhist koan, or mind puzzle, but to me it's just a stupid question that any Zen teacher should be ashamed to ask. In the first place, everything that exists has Buddha nature, according to Zen master Shunryu Suzuki, who should know. But even if everything didn't have Buddha nature,—how about unconditional love?

Mothers are supposed to love their children unconditionally, and maybe some do. But does a dog ever ask, How come you never visit? When you leave your dog, she will cry. When you come home, she will wag her tail. No questions asked. That was then, this is now. For her it is always now. She needs no meditation teacher, no practice. She is already perfectly herself, in the moment.

A few days before Thanksgiving, having almost finished writing her story, I knew it was time to speak with Daisy again. I wanted to know how she was doing, and how she liked her book. I was especially curious to know if she would be able to tell me anything about her life before she came to me. I didn't know whether this was possible, whether it was too far back or might be too painful for her. I tried to have no expectations. It would be enough just to be able to talk with her again.

I made a list of questions and called Karen Beth.

Karen had a new cat, Maizy; but thereby hangs a truly remarkable tale. After Perdita died Karen and Gail De Sciose had both contacted her and learned that Perdita was planning to reincarnate, again as a cat. To Karen, Perdita described the house where she would be born: sliding glass doors leading onto a deck, wood stove, stone fire place. Through a friend Karen heard of a cat who was pregnant, went to the house, and found everything as Perdita had described it. When the kittens were born Karen had no trouble telling which one was Perdita.

In Buddhist Tibet, after the Dalai Lama dies, he communicates the time and place of his next incarnation through oracles. The Karmapa, the spiritual leader of the Kagyu lineage, leaves detailed written instructions about his rebirth that may even include the parents' name.

I was aware of this ancient tradition in a culture that takes reincarnation for granted. But this was the first time I had heard of an animal not only reincarnating but telling her person where to find her. When I told Karen's story to a Buddhist friend, she smiled and said, "Karmapa Kitty!"

I asked Karen if she would be willing to contact Daisy again. She agreed. As before, I gave her my list of questions. She called me back with the answers and later sent me her written report:

FOURTH CONVERSATION WITH DAISY, NOVEMBER 23, 1998

I contacted Daisy, told her who I was, and asked her if she'd like to speak to me.

Daisy: Yes, I would. I have been waiting for someone to contact me.

I see her very happy, wagging her tail.

Karen: Where are you?

Daisy: I am not in a body, yet I live.

Karen: Is there anything you'd like to say to Helen?

Daisy: Tell her I miss her and want her to be happy. I love her so much. While I was on Earth she was my friend, my companion. I'm not ready to go back to Earth, but some day. Helen spoke so gently. I love her voice. I am happy, I am whole. I live in pleasure without pain. I am close to the Earth, not far. I visit Helen in her dreams.

Karen: Do you know that Helen is writing your story?

Daisy: Yes, she told me she would.

Karen: Do you like it?

Daisy: Helen has not read it to me.

Karen: Do you want her to?

Daisy: Yes, that would please me very much.

I asked her about her life before coming to Helen. I saw human feet stamping near Daisy, saw her looking through garbage, hungry, animals chasing her.

Daisy: Helen came for me in a van.

I wasn't sure I was really seeing this. Then Daisy looked deep into my eyes. I felt her paws in my hands. I started crying. I was not getting any images or words, but I was feeling deep sorrow and despair.

I think she might have been trying to show me more pictures, but what I was able to pick up was the feeling. It seemed like she appreciated being able to share this with someone.

[Helen's note: I had asked Karen to ask Daisy if she minded if in this book I talked about my experiences communicating with other animals besides her.]

Karen: How do you feel about other animals being in the book?

Daisy: The more people who know we are not just "dumb animals," the better. Humans are a strange lot. They need to be healed. Tell Helen to think about me before she goes to sleep. I will contact her in her dreams. It was a joy to be with her in life.

I had assumed that since Daisy was in spirit, she not only knew I was writing her story but had probably been helping me with it. Maybe she had. But now I knew she wanted me to read it to her.

The next morning I made a fire in the fireplace and turned off the ringer on the phone. I put the chain with Daisy's license tag on it around my neck and put my favorite picture of her on a table where I could look into her eyes as I read. I got the latest version of her story and sat down in Mother's rocker in front of the fire. I closed my eyes, took some deep breaths, and mentally called my dog. Then I opened my eyes and started to read.

I read aloud, and as I read, I could feel Daisy's energy. I could also feel when something wasn't quite right, and wrote in some changes as we went along. Daisy was my first editor!

Before Daisy died I spoke with her through Gail De Sciose, Karen Beth, and Ginny Debbink. To learn more about this phenomenon, which Penelope Smith calls "interspecies telepathic communication," I decided to contact each of the three after Daisy died with a similar list of questions. I hoped this would not be confusing for Daisy and that she wouldn't mind answering some of the same questions again. I spoke with Gail on the first day of December.

FIFTH CONVERSATION WITH DAISY, DECEMBER 1, 1998

> Gail said, "I told Daisy, I am one of three animal communicators who have spoken with you. Does it bother you to speak to Helen through three different people? Daisy says, *Please tell my dear Helen that it is never a bother to speak with her*."
>
> I asked, "Does Daisy have anything to say to me?"
>
> "Daisy knows that the Earth is turning and the seasons are changing. She remembers that you used to like walking together when the leaves were turning and when the snow was falling. She says, now that she is not in a body, she hopes you will do

this for yourself because in doing it for yourself, you do it for her, too."

"She is a high being. She is my teacher. Please tell her that."

"Daisy says the teaching and the healing did not just go one way. You and she supported each other on your journey. She can continue to do this. You can call on her. Daisy admires you very much."

"Where is Daisy now and how does she feel?"

"Daisy says, *I feel strong in my body. Watch me run!* I see her outside, by water, watching ducks or geese, not as a predator."

"Was there water where she lived as a puppy?"

"She lived in a very rural area, a farm, with other animals. There was a pond, with geese. She and her littermates were born there. She was outdoors a lot. There was not much bonding with humans."

"Was there an owner after that?"

"This place sold farm products. When Daisy was four or five months old, a young man with longish hair came to buy apples, and took her away. She was not with him too long, maybe a year. He took her everywhere in the back of his truck. She slept near him. He liked music, had a guitar. He took good care of her. Then he went away, and no one took care of her. She was wandering."

"Daisy came to me in October of 1983. Does

she know what time of year she was born?"

"The young man came in the fall, when the farm was selling apples. If she was four or five months old in September or October then she would have been born in the spring, April or May." (This tallies with the vet's estimate that she was a year and a half old when I adopted her in October.)

"Does she remember meeting me?"

"She is showing me a place with cages, lots of other animals. People would come by and not look at her. She wasn't sure whether she wanted them to look at her or not. She did not want to be with the wrong person. It was a terrible place with concrete floors. She was afraid. Time was running out—animals are conscious of this—and she had to find the right person. She says that you were looking with your heart. She had had a rough time, losing first her mother and littermates, and then the young man."

"Did she hear me read the story to her? Does she like it?"

"She likes it very much."

"Is there something she would like to say to our readers?"

"Daisy says that when humans look at an animal, it is much more important not to see the physical being in front of you, but to open your heart. Daisy thinks that if you had not looked with your heart she would not have found you, because she thinks she wasn't too striking looking."

"Tell her she was elegant! She received many compliments."

"She says, *So why were people always asking you what I was?*"

"Daisy told Karen she would come to me in dreams, that I should think about her before I go to sleep. I've been doing this. At first, nothing happened. Then I dreamed I was trying to phone my friend John who died in 1982. But when I woke up I realized the number I was dialing in the dream was the number of Daisy's vet. The next night I dreamed about Max, the German shepherd who lived with us for a while. The night after that I dreamed about a little black dog who looked a lot like Daisy but did not have her white markings. Please ask Daisy if that little black dog in my dream was her, and how I can help her come to me in my dreams?"

"Daisy says, *Does Helen know that everyone is a part of everyone else? Every time Helen dreams about a dog, a part of that is me. Before Helen goes to sleep at night, she could set up a little altar with my picture, light a candle, and ask me to come.*"

"Does Daisy understand my words, or does she read my thoughts without words? Does she hear me when I talk to her?"

Gail laughed. "Daisy says, *Helen goes along as if I'm right there, and I'm not right there!*"

This hurt my feelings a little. "But she said she would never be far from me."

"She means she's not there in a body. She says, *You don't have to talk out loud.* I get a feeling of great peacefulness from Daisy. She is moving around, running, but it is not agitated."

"Gail, when you do this work do you get words, or do you get images which you then translate into words?"

"I get words, images, and emotions. Some animal communicators get only one. I'm lucky, I get all three."

"Does Daisy ever go back to the place in Connecticut where we used to live?"

"Daisy says, *That is just a place. It was once important to me because the person I love was there. Now it is just a place.* She has created parts of it to visit in spirit. She doesn't have to be there."

"I was going to take some of her ashes and scatter them there, but I guess she doesn't care where her ashes are."

"No, she doesn't."

"Please tell her how much I love her."

"Daisy says, *And I love her.* She hopes you can feel that. She says to tell you, I very much enjoy being a dog. If I come back, I'll be a dog again."

I wanted to finish Daisy's story in time for Christmas. On the Winter Solstice I spoke with Daisy again through Ginny Debbink.

SIXTH CONVERSATION WITH DAISY, DECEMBER 21, 1998

When Ginny contacted her Daisy commented, *"This is getting to be a habit!"*

"Does Daisy have anything to say to me?"

"Daisy says she's very proud of you for the work you've been doing. She sees that you have a tendency to be too serious. She asks me to remind you that it's supposed to be fun."

I assured her, "I have enjoyed it. How is she

feeling now? What is it like where she is?"

"She says she feels wonderful. She feels light. She says, *It's always warm here.*"

Again I was reminded of the times she was homeless in all weathers, and of her passion for lying in the sun even on the hottest days.

"I don't understand this, but Daisy says, *Sometimes I'm in the blue and sometimes I'm in the green.*"

This made perfect sense to me. Blue and green are the colors of planet Earth and her atmosphere. "I think she meant, sometimes I'm floating through the air, sometimes I'm on the ground.

"Does Daisy have anything she wants to say to our readers?"

"She says, *It is important that they know to go softly.*"

I asked if she could explain what she meant.

"She says, *It's about understanding and valuing every step. Stepping gently, and at the same time, never taking it too seriously. Understanding is not as important as valuing. Value every step.*"

This just bowled me over. Not only is this the central teaching of Zen master Thich Nhat Hanh; it is also the message of the book I had been reading for the past week, Robert Pirsig's *Zen and the Art of Motorcycle Maintenance*. It was as if Daisy, through

Ginny, lifted this idea right out of my head.

But on an even more profound level, that was the great lesson that Daisy had taught me while she was here with me on Earth. Those days I didn't want to get out of bed and walk her, but did, I remembered as we walked, bound together by that leash, in sun or rain or snow, that every step was precious. And here she was, my dog with Buddha nature, reminding me again of the truth we always forget: Every step is precious.

"Tell her, I sit at her feet. Has she had other lives on planet Earth, either as a dog or as some other species?"

"Daisy says that she is a very old soul. She has had many lives, frequently as a dog. She likes the life of a dog, and she is good at being a dog. She says that every incarnation she has had as an animal, she has been black and white."

"Has she had lives on other planets?"

"Yes, many lives, many places."

"Has she ever had lives as a human?"

"Yes, several times. The life she remembers most vividly was that of a young peasant girl in some foreign country where it was very hot, possibly in Asia. She had worked hard and died young, but it had been a good life. She had had no children, had been a healer or midwife."

"Does she plan to come back to Earth some time, and if so, what form would she choose?"

"She says, *Yes, my journey is not done. I still have many lessons to learn, especially patience and forgiveness. These lessons are easier to learn in a dog's body. By and large, dogs are better at patience and forgiveness than people, but we are still not perfect.*"

"How does it feel not to have a body? Was it scary at first, or did she remember being without a body before?"

"Daisy says, *It is not necessary to remember. It is not scary. There are no boundaries or limits. To know what it feels like to be without a body, think of joy. Imagine having arms large enough to take in the whole sunset. Multiply that beyond imagination, and that's how it feels. Without a body there is no fear.*"

"Are there other beings where she is, animals or humans or both?"

Ginny chuckled at Daisy's language. "She says, *There are multitudes, and they are everyone. The question, animal or human, is meaningless here. We are all pure spirit.*"

"If she decides to come back as a dog, will she please let me know so that I can try to find her?"

"Right now you are the spirit to whom she is most closely tied. She says to remember what she said about valuing being more important than understanding. It is her hope to be with you again. You must remember that we are all one family, but we all have smaller groups of souls with whom we have traveled in many lifetimes and with whom we have many obligations, debts, and lessons. When she comes to a clear understanding of her own path of growth, that will be her time of decision. She says, *Lessons we don't choose. Our path we choose.* I don't understand this. Do you?"

"I think so. There are certain lessons we are all here to learn, like patience and forgiveness. The need to learn those lessons is a given. But how we go about learning them is up to us."

"She says, *Yes, there are things that we each continue to get wrong.*"

"Does Daisy understand my words or does she read my thoughts without words? Is it different now than when she was in a body?"

"Daisy says, *Neither words nor thoughts. I listen to her on an emotional level. I read her level of truth and intent. Yes, it is different now. There is more clarity, and there is a deepening. The body gets in the way of communication.*"

I asked Ginny, "When you do this work, do you get images, words, or emotions, or all three?"

"All three, and sometimes sense impressions as well. It is a knowing," she said.

"I send Daisy oceans of love."

"Daisy says she receives and returns the love. She says to tell you, *When you do that—when you receive and return love—you are close to the truth.*"

Animal communication seems mysterious and yet natural. Those who do this work say that we all have the ability, but few of us are aware that we have it because our culture does not encourage its development. The secret of communication with animals, like communication with certain humans who are hard to reach, could be as simple as believing it to be possible and taking the time and trouble to do it.

I think about animal communication a lot: what a gift it was for me to speak with Daisy before she died, what a wonderful work it is these animal communicators do, and how important it is that humans know that this kind of communication with other species is possible. And what a comfort it is to me now to know that Daisy is spirit, that she will always be with me, and that I can speak with her again.

To people who have never experienced this phenomenon I say, Imagine what it would be like if for years you lived with someone you loved dearly who spoke a different language. Then imagine that at the end of her life, an interpreter appeared, and you and she were able, for the first time, to communicate in words. That's how it felt with Daisy.

My mother had a living will and had often expressed to me her wish not to be kept alive artificially. Toward the end of her life she had trouble swallowing, and during one of her hospitalizations she was given a swallowing test which the hospital said she failed. The doctor asked my permission to put a tube in her stomach. Because Mother and I had spoken of these things I was able, after much soul searching and consulting with my brother and nieces, to tell the doctor, No tube. (In fact, once home, my mother could swallow again.) It was an agonizing decision, but at least I had Mother's written and spoken words to go on.

With Daisy, I had no words—until I called upon Gail and Karen and Ginny.

Were the words that came through the animal communicators really Daisy's words, or were they words that these women found to translate the pictures, feelings, and ideas they received from my dog? Were they really channeling Daisy, or their own inner wisdom? I don't know, and I don't think it matters. What matters to me is that the words carried a precise personality and a boundless love that I recognized as real, and contained a world view that has, for me, the ring of truth about it.

Daisy taught me so much, and some of the words that came through the three women were lovely in and of themselves: *Look with your heart. Value every step.* I decided to cull my favorites from the six sessions and weave them together to form a whole.

Beyond

The Sanskrit word *sutra* means "a thread on which jewels are strung." In Hinduism and Buddhism the word is used to refer to a collection of aphorisms or teachings. In *The Lotus Sutra*, for example, the Buddha dispensed jewels of wisdom to his disciples. With no disrespect to the Buddha, I'd like to offer *The Daisy Sutra*: a string of pearls that are no less valuable to me because they came as a gift from my dog:

The Daisy Sutra

I am not in a body, yet I live.

Here, there are no boundaries or limits. To know what it feels like to be without a body, think of joy. Imagine having arms large enough to take in the whole sunset. Multiply that beyond imagination, that's how it feels. Without a body there is no fear.

Here, there are multitudes. We are all pure spirit.

My earthly home is just a place. It was once important to me because the person I love was there, now it is just a place. I have created parts of it to visit in spirit.

My journey is not done. I still have many lessons to learn, especially patience and forgiveness. These lessons are easier to learn in a dog's body. By and large, dogs are better at patience and forgiveness than people, but we are still not perfect.

It is my hope to be with you again. We are all one family, but we all have smaller groups of souls with whom we have traveled in many lifetimes and with whom we have many obligations, debts, and lessons. When I come to a clear understanding of my own path of growth, that will be my time of decision. Lessons we don't choose; our path we choose.

The more people who know we are not just "dumb animals," the better. Humans are a strange lot. They need to be healed.

When humans look at an animal it is much more important not to see the physical being in front of you but to open your heart.

I receive and return your love. When you do that, when you receive and return love, you are close to the truth. Everyone is a part of everyone else.

It is important that humans know to go softly. It's about understanding and valuing every step. Stepping gently, and at the same time never taking it too seriously. Understanding is not as important as valuing. Value every step.

The Earth is turning and the seasons are changing. We used to like walking together when the leaves were turning and when the snow was falling. Now that I am not in a body, I hope you will do this for yourself because in doing it for yourself, you do it for me, too.

Interview with an Animal Communicator

I wrote the story of Daisy's life because I had to. Losing her was bad enough, but the thought of losing my memories of her, of forgetting a single detail, was intolerable. I needed to get it all down, and I needed to share it, to honor her, and in honoring her to honor all dogs.

Painful as it was to lose her, Daisy's death brought me a great gift: the discovery of animal communication. Her death was a doorway into a whole new world, a world in which we humans are no longer so alone.

The decision to have a beloved pet put to sleep is agonizing. The spiritual perspective and practical information I received in my sessions with Gail, Ginny, and Karen helped me to ride the emotional roller coaster of grief with my dog still beside me. Animal communication breaks down the walls between species, but it does more than that: by showing us that animals are spirit, it also breaks down the walls between the living and the dead.

Now Daisy's story took on another dimension and another purpose: I wanted to share what I had learned about animal communication. I wanted people to know that we can talk to the animals and they can talk to us, that animals are our teachers. I had had a few experiences of communicating with animals on my own, but I wanted to know more about it. In a skeptical age like ours, how does one come to this work?

I spoke with Gail De Sciose one day in March of 1999 in the sunny Manhattan apartment she shares with her husband, Joseph. No animals were physically present, but as Gail told me stories about her work with animals, the room filled up with their wise, comical, and loving spirits. (Note: To distinguish our voices in the interview that follows, the words in italics are mine.)

When did you start doing this work of animal communication, and how did you get started?

I am not one of those people who communicated with animals from the time I was a child. I always loved animals, but unfortunately I was very allergic to them, so I had to keep my distance. I read books about them and just loved to see them. I tried to be around them whenever I could.

In 1988, as an adult, I started volunteering at the ASPCA animal shelter here in New York City. I counseled people who wanted to adopt, I helped socialize frightened animals, I walked the dogs and played with the cats. It was while I was trying to socialize some of the frightened animals that very strange things were happening. I would get these crazy ideas about how to help the animals, and they wouldn't necessarily be logical or even particularly safe ideas, but in many cases I would act on them and they always seemed to work. For example, letting a snarling dog out of a cage and letting him come to me, without my being harmed.

I would also take animals down into the basement of the ASPCA. I would sit on the floor and meditate with the animal on my lap and they would calm down. I knew something rather extraordinary was happening but I didn't know what it was. I went on like that for a few years.

In 1994 a friend told me about animal communication and Penelope Smith, who's quite well known in this field. I was really very skeptical, but I was intrigued by the whole idea. I called about classes, and that summer I took a workshop with Penelope up at Spring Farm, a rescue farm near Utica, New York, run by a communicator named Dawn Hayman.

It was my first class in animal communication, and I went up there thinking, Why am I here? Why am I doing this? I don't really believe this. But as soon as the class ended, I was communicating with animals. I couldn't do it too well during the class because I was very nervous about it and put a lot of pressure on myself. In retrospect that was actually a good thing, because now that I teach, that experience helps me to understand the doubts and hesitations of my students.

The class consisted of a Friday night lecture and a half day each on Saturday and Sunday. At the end of class on Saturday Penelope gave us an assignment, to just observe an animal. She said, "You don't even have to communicate with it, just observe it."

My husband and I were staying in a bed and breakfast, and the only animal I could see was a spider who was building a web in some flowers that were in our room. It was the last morning of the class, and I was doing my stretching exercises and my yoga. I kept watching the spider every time I turned my head. I had a very strong feeling that it was a female, and I could see the precision and the great orderliness of her work. I was really in awe of the job she was doing. Then all of a sudden I saw the flowers, but they were huge and they were in color. I was actually seeing what she was seeing!

Oh, you were seeing as her!

Yes, I was seeing through her eyes, essentially. I thought that was fairly extraordinary. At class that Sunday Penelope asked about our experiences.

I said, "Well, I was observing a spider but I know I made this up, because I saw the flowers in color and animals don't see in color."

I remember Penelope laughing. She said, "You've obviously been watching too many *National Geographic* specials because they *do* see in color, and I'm sure you saw what the spider saw."

That was my first conscious communication. I really think it was incredible that it was with a spider. People don't think you can communicate with animals like spiders and snakes.

I came back home intending mainly to help the shelter animals, but word got around. People would say, "You know, my dog does this really weird thing. Would you ask him why he's doing it?" or, "My aunt's cat is not eating the food that she's eaten all her life. Can you help?" So I started doing more communicating.

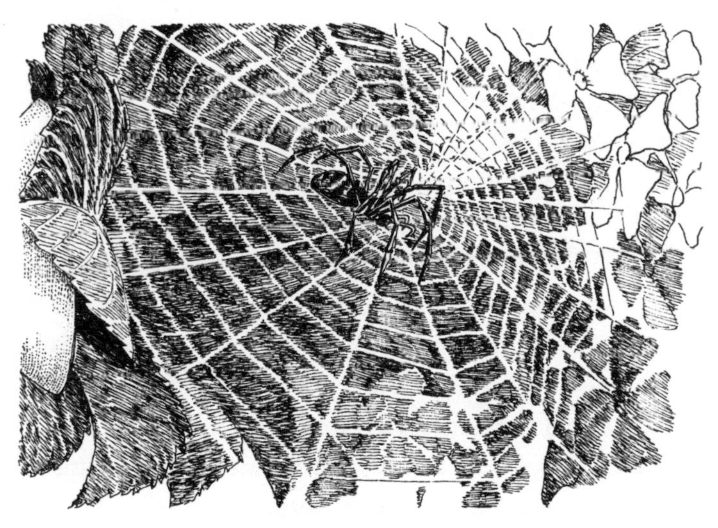

In the beginning I couldn't do it on the phone, the way you and Daisy and I did it. That was too much pressure for me. So I would take down all of the person's questions for their animal and then I would go off by myself and meditate for half an hour. Then I would start to have the dialogue with the animal.

So the animal didn't have to be there.

No, the animal is very rarely in my proximity, even now. I work very well telepathically. Even when I'm sitting face to face with an animal I still don't necessarily verbalize what you want me to ask. I do it telepathically, silently, from my mind to the animal's mind.

Initially, I would write down all the dialogue—what I said and what the animal said. I have notebooks full of these transcripts. The problem with doing it that way, I learned, was that if the animal brought something up that you as the person might have wanted to follow up on, I wouldn't know that, so I would just go on to the next question. Now when we do it on the telephone you can take the conversation in any direction you want to, because you know your animal.

When an animal is telling me something, I have no idea if it's the truth or not. I have to get the confirmation from you. It's a very interesting process because it's kind of a leap of faith every single time. I never know when that phone rings who's there, what it's going to cover, how articulate or inarticulate the animal may be or what emotions I'm going to be dealing with. It's always an adventure.

I've been doing animal communication for five years now—consciously, anyway. The first year I did it I worked with about 140 animals. In the summer of 1995 I went to California and took advanced training with Penelope. After that I became a professional because communicating was taking a lot of my time and energy. I've attended holistic seminars and worked with holistic vets and so on. At this point I've dealt with thousands of animals and about fifteen hundred people. Many people use me over and over again.

Do you have a daily meditation practice?

Yes. It's the first thing I do when I get up in the morning. It is through meditation that I became an animal communicator. I also have a daily prayer regimen which has to do with Reiki [an ancient Tibetan healing technique that came to the west from Japan]. I'm a Reiki Level Two practitioner. I keep a list of all the people and animals I want to pray for.

You said that when you first took Penelope's class you were skeptical. Did your skepticism completely disappear that weekend, or was it a gradual process?

I had a couple of other experiences that weekend. I communicated with a cat at our inn and she told me things which her person later verified. I communicated with a horse in a pasture and very definitely knew that I had his attention; and then the spider. When I came home, I had some very extraordinary experiences. Within the first month or two of my learning to do this I was communicating on a very deep level about life-and-death issues with animals.

At first I didn't believe that you could communicate with animals in spirit, as Penelope had maintained. I thought it was pretty extraordinary that we could talk to the ones who are living.

But there was a dog at the shelter who was a special friend of mine. He was very ill and had to be euthanized. He came to me the night he was euthanized. He is one of the very few animals who have just come to me. Most animals I go and look for.

I have strong boundaries, living in New York City, and I also have real ethics issues about invading people's and animals' boundaries. But this dog from the shelter did come to me.

It was a Thursday night about 8:30. I was at a group meditation program with about 250 people, so it was very powerful. All of a sudden, the dog was there.

Had he passed over at the time?

Yes. He came to me to show me that he was fine, that he was happy, that he was physically whole. He was running in a field. The next day when I went to the shelter I asked about the dog. They said that he was euthanized about 8:30 the night before.

That reminds me of my experience with Max, when I heard the bark over the telephone, and the person I was talking to said, "What bark?" And it turned out to be the time that Max was put down. I was in Kingston Hospital at the time, and there weren't any dogs in the building!

Yes, your experience illustrates my belief that many people do have instances of animal communication. They simply don't recognize them as such at the time.

I think my skepticism disappeared very quickly, probably within the first couple of months. After seeing the results that people were having in Penelope's class, I was convinced that interspecies communication was a real thing. Penelope says that she's been doing this all of her life and she does it effortlessly.

I don't doubt it at all any more, and the animals are always teaching me. It's an ongoing learning experience. I also view it very much as a spiritual practice because it does put you in touch with the fact that we are all connected. The animals know this so beautifully and they express it. They are not afraid to leave their bodies because they know about their souls being immortal and they understand about reincarnation. Some of those things I had previously said I believed in, but nobody ever put me to the test. Now it's my daily experience in dealing with the animals.

When you realized that this phenomenon was real, did that challenge your previously held world view, or did it just confirm it? Were you open to the idea?

Yes, I was very open to it because of my meditation practice.

What kind of religious upbringing, if any, did you have?

Technically, I was brought up Jewish. My mother was Jewish. She divorced my father and married a Catholic, but we continued to go to temple. I wasn't super-religious, but I always believed in God and in some kind of a spiritual connection. What really broke everything open for me was meditation. I started to see the great possibilities of things through doing meditation. And through my practice, my life changed dramatically. At this point I don't attend regular religious services, but I do feel that my spiritual practice is the essence of my religion.

So when you started, you saw fairly quickly that animal communication was real, but you were still surprised that you could communicate with an animal after it died.

Yes. But once I had that experience with the dog at the shelter, I knew this was real. It was a great gift he gave me to come to me that way after he died. After that, I started communicating with animals in spirit, and there was no looking back. I've never doubted it since. And I've never had any problems with doing it. It's not weird. It's very natural.

I was brought up Christian, and neither Christianity nor Judaism, for the most part nowadays, accepts reincarnation. In 500 A.D. the church fathers condemned the doctrine and removed most references to it from the Bible. They wanted the faithful to focus on this lifetime and on salvation. But of course reincarnation is taken for granted in Asia. I had an experience when I was fourteen that opened me to reincarnation. What was your feeling about it before you came to the animal work?

I thought it was possible—again through my spiritual practice. And it's been confirmed for me by the animals. In my five years of doing this work I know of at least ten animals who people are very certain have come back to them in another animal body. In some of the cases the animals were able to tell us while they were in spirit how to find them, and we were able to find them.

As with Karen's cat. Is that one of the ten?

Yes. And then there was a cat in Germany who told us how we would find him, and a dog up in Canada. Some of the others were a little more casual than that, but those three told us very specifically where to find them.

Was it the precision of the instructions that the animal gave that convinced you, or was it the person's certainty that this was the same personality?

Both. The cat in Germany told us he was going to be an orange tabby cat. Ingrid, his person, spent a long time looking for an orange tabby cat in Germany and couldn't find one. They're apparently not very common there. Then she went on holiday to Italy and was staying at a farm. The farmer's cat had had a litter of three kittens, and one of them was a little orange and white cat who came in front of Ingrid and just looked at her. The farmer had told Ingrid that this mother cat never showed her kittens to anyone because the farmer had a tendency to drown them. But in this case the mother cat allowed her kittens to be seen.

Ingrid called me from a phone booth in Italy and said, "Do you think that this is my cat, Moritz?" So I tuned in to this little kitten who was just a couple of weeks old. I said,

"Are you Moritz?" And he said to me—it was so cute, I remember the exact words—"It is I, Moritz. I am returned!" Ingrid brought him back to her home in Germany. She brought the mother and the other two kittens, too, because they were too young to be separated.

She brought the whole kit and caboodle!

Right. And Moritz was Moritz. When he got back to their house, he was doing the same kinds of things that the other Moritz had done. She named him Moritz again.

The dog in Canada told us, again with great precision, the time of year that he would be born, that he would be born into a litter of puppies, that the litter would be born to a friend of the person, and that his coloring would be very different from the other puppies. Everything he said was true, and he also acts the way his former self did.

But I will tell you that there are some communicators who say they never encounter cases of animals reincarnating.

Maybe it's their world view, or maybe the cases involving reincarnation don't come to them.

I do believe that the people who come to me are the ones who are supposed to come. That's what I pray for every day, that only those people and animals I can truly help will come.

There are stages of belief here. It's as if things get harder and harder for a true skeptic to believe: first, communicating with animals while they're alive; then, communicating with them after they die; then, having the animal plan its next incarnation in a precise way that gets confirmed; and finally, we come to the idea that an animal—a non-human animal—could have a human incarnation.

On her tape Animal Death: A Spiritual Journey, *Penelope Smith tells about a leopard she saw in a zoo, pacing and yowling pitifully in a small cage. She asked him why he was there and he showed her images of his past life as a man who poached animals illegally, including leopards.*

The leopard had been human and was doing this life as a caged animal to pay off that karma.

I can only tell you that my experience, at least so far, is that I view everyone as an individual, whether they're a cat or a worm or a snake. I once tuned in to a cat who said, "Convince me as to why it is in my best interest to talk to you."

That sounds like a cat!

I asked that cat, "Have you ever been a human being?" He said, "Do you think I would have this kind of vocabulary if I had only ever been a cat?" I thought that was pretty amazing.

I have had people ask me to ask their animals to show them past lives and to tell them whether they have been together before. Sometimes the person and the animal were both people together, sometimes they were two animals together, or sometimes the animal was the human and the person was the animal. It seems to go all over the board, which I had a little trouble with, initially. I had believed that the soul starts at ground zero and evolves into higher and higher life forms. Going back and forth from a human to an animal body didn't accord with my previous idea of evolution.

But I came to realize that there's no duality, there's no higher and lower. I think sometimes you have to be in a certain kind of body to get the lessons that you need in that life. If you're an antelope being chased by a lion, then that's the lesson that you need to learn. I now feel that we do go back and forth and all around, and I try to honor and respect everybody, whether or not they're a so-called intelligent life form. People tend to think that intelligence means acting more like *we* act, that the intelligent animals are the apes or the dogs who will answer your call and do tricks, and the less intelligent animals don't do that. But that has nothing to do with intelligence.

It's an arrogant attitude.

It's an anthropocentric, egocentric attitude.

Do you pick up things from people as well as animals?

I pick up things from people in conjunction with their animals. Going back to the boundary issue, I don't walk around open twenty-four hours a day, seven days a week, three hundred and sixty-five days a year. Let's say I'm doing a communication with you and Daisy. If you want me to ask Daisy what she thinks about things that are going on in your life together, you won't even have to tell me what those things are. But it seems to be reflected through the animal's eyes.

I always ask permission to speak to an animal. I do a prayer before every session, unless it's a dire emergency—

somebody's dying on the other end of the phone—in which case I just say, "God help me to do this," and then I address the animal. For example, I'll say, "Daisy, this is Gail, we talked before. I want to know if you'll speak with me because Helen would like me to talk with you." I wait to hear, and if an animal said "No," then I would say, "I understand, that's fine, you don't have to talk to me."

That's only happened twice so far, and in each case I honored the wishes of the animal. In one case it was a pony in a pasture. He said, "Get along, don't bother me," and I said, "Fine, I understand." I started walking on down the fence line to see some other animals. He came right over to me because I had paid attention to him and honored his wish, and then he did talk to me.

Was there a human involved with the other animal who refused?

Not directly, no. The animals were voles. When I deal with a pack of animals, I usually ask to speak to what Penelope Smith refers to as the oversoul of the animal. Once when I had been trying to contact a female deer, this deep booming male voice showed up. I said, "Who are you?" He said, "I am the one who is in charge of all the deer." In those cases where it's not a domesticated animal and it's an animal group, I will speak to the one who is in charge.

These particular voles were under someone's yard, and they were destroying all the shrubbery. I talked to them twice. I told them that they were going to be poisoned and that it

wasn't in their best interest to stay where they were. They basically said they didn't care, they were hungry, and if they died, they would just get into another body and come back, no big deal. The next time I tried to speak with them, to tell them that poisoning day was upon them, they totally tuned me out. They wouldn't respond to me in any way. And of course some of them were poisoned, but that was their choice.

Are you ever in a situation where you pick up animals communicating with other animals?

Yes. In fact, I'll tell you a very funny story. I have a long-time client who has five cats. This client happened to call a psychic—not me, I don't consider myself a psychic—about a personal problem. At the end of the session the psychic said to her, "There's a cat in your house who wants to tell us something." So they let the cat come through and the cat said, "There's a cat here who is sick." She didn't indicate who the cat was, but my client said at that point, "I know how we can find out." She got off that phone call, called me, and we started talking to the cat who had told us about the sick cat. She said that a gray tabby cat in the house had something wrong inside of her that was very bad. So I tuned in to the gray tabby cat, and it turned out she had a serious kidney problem which wasn't manifesting in any physical symptoms.

My client took the sick cat to the veterinarian and said, "I want you to check this cat. I think something's really wrong with her kidneys. I want you to take a urine culture." The vet said, "Okay, leave her here for the day. I'll check her out, and you can pick her up tonight." The day came and went, and my client went back to get the cat. The vet hadn't bothered to culture her urine because the cat was behaving normally. So the client said, "Look, just humor me, culture her urine." When he went to collect the urine, it was full of blood. He said, "How did you know this? Was she misbehaving, or acting in a strange way?" My client was bold enough to say, "No, one of my other cats told a psychic, and then I called an animal communicator."

She was very brave!

I'll say! Because the vet knew her, he humored her.

A diagnostician of a cat!

Right. I have another client, who calls me fairly regularly, who has a cat who kind of watches over everybody else in the household, the cats and the dogs. This cat will tell us if something's wrong.

Speaking of animals communicating with other animals, I just remembered a story my friend John Button once told me about a farmer who buried a litter of newborn kittens alive. The mother cat found the place where her kittens were buried and began frantically digging. But the hole was too deep for her, and pretty soon she was near exhaustion. So, John said, she went and got the dog, and brought him to the hole she had started, and the dog dug through to the kittens. They were still alive. The farmer was so impressed that he let the kittens live.

Getting back to the subject of animals reincarnating as humans and vice versa—transmigration—does that come up often?

No, because not that many people go there with their communications. When I tune in to an animal, I get a general sense of the animal's personality in this particular incarnation. If the client just wanted me to ask the animal, "How do you feel? Do you like your food? What can I do to make you happy?" that's as far as it would go. It's unusual for an animal to start talking about past lives without being prompted. Every once in a while I get people who say, "Ask him if he's ever been with me before. Have we ever had lives together?" Then they'll go there. But the animals don't bring it up as a rule. It's just sort of a fact for them, it's nothing extraordinary.

It's not a big deal, and they're also very much in the present. That's one of their great gifts to us.

Chris Griscom, who wrote *Soul Bodies*, does a lot of work with reincarnation. I once asked her whether she had ever had the experience of animals crossing over and being human and humans being animals. She said she hadn't as far as she knew. But she did know of animals who would reincarnate over and over again to follow their people, in a kind of parallel path.

Dannion Brinkley and Betty Eadie have both written books about their near-death experiences. I asked each of them if, when they were making their transitions through the tunnel of light, they had any awareness of animals being with them, and they both said they did. They heard them and Dannion said that he saw them, too.

I just had a woman call me from down south. She said, "In the churches down here they tell you that animals have no souls. I don't believe that."

A lot of people can become very distraught because of their religious teachings on that subject. They'll go to church, and they'll hear that an animal doesn't have a soul and can't go to heaven. I don't want to make any religion wrong, but that is not my experience. When I give a lecture or an interview, I'm speaking not from what I've read, not from what I've heard, but from my experience. My experience keeps accumulating, and the animals keep refining it for me. My experience is that animals do have consciousness, do have souls, do go on to an afterlife, and can come back.

You once told me that Penelope used to encounter ridicule, that people can get really upset and very negative about animal communication. Have you ever come up against real hostility?

Yes. I remember Penelope saying to us at an Advanced II

class, "You know, you people have it easy. I was out there for the last twenty years telling everyone this is real, and people laughed at me. Now it's much more widely accepted."

After I was on *Coast to Coast* [the Art Bell radio talk show] the first time, a man called me and said, "I'm from Canada, and I can tell that you're from the United States and that you're a space cadet, because no one in Canada would say that they could talk to animals," and he slammed down the phone. He was so incensed that he had to call me and say that.

A few years ago I did a communication with carriage horses in Central Park for a television program called *Nature*. I was fairly nervous because it was my first TV appearance. The horse I spoke to was very enthusiastic and said, "Let's go, let's get in the park." I couldn't believe how excited he was to go do his job.

Really?

I would not have thought that. I feel very bad about the way the carriage horses are usually treated.

What time of year was it?

January. It was freezing. In fact it even started snowing in the course of the communication I was doing. But the horse was very vigorous and spunky. After the filming session the driver of the carriage came up to me and said, "All right, if you can communicate with this horse, tell me where he came from. "I said, "Do you know where he came from?" He said, "Of course I know where he came from. I bought him." So I asked the horse about his life before he came to be with this man.

What the horse showed me and told me about was being on a working farm. There was a young boy who was his good friend who would come and bring him special treats. Apparently the boy would ride him even though he was a work-type horse. So I told the man, "He came from a working farm and he worked hard but this young boy was his friend and brought him treats."

The man said, "You're wrong. He came from an Amish farm. He was a plow horse and they worked him very hard. I recognized that he was doing a great job and he had all this spirit, so I bought him and I brought him here to New York."

He said, "If horses were smart they wouldn't be working like this. They've never written symphonies," they've never done this, they've never done that. I said, "Fine, I'm wrong." But later I thought, the reality for this horse was that he chose to focus on the relationship he had with this boy, not on how hard his life must have been on that farm.

The stories don't contradict each other. They complement each other.

I agree. And you know, just because a horse can't write a symphony. . . .We can't fly without the aid of machines. Every creature has a purpose and a body adapted to that purpose. Animals are miraculous and marvelous and we learn from them. We are emulating them by doing the things that we're doing. We just happen to have a brain that is very complex and has symbolic language and can figure things out. But animals can figure things out too.

What do you say to skeptics? What do you say to people who love animals deeply, but just don't think that this kind of communication is possible?

I can only tell them the place I started from. I loved animals deeply, I didn't believe this was real, I went to check it out. It became my experience, which has been revalidated constantly over the past five years with thousands of animals and people. If someone would just allow the possibility that this can be true, it could just break their former concepts wide open—which is why some of them won't do it, because it's too frightening.

If you grow up having a certain view of the world, and somebody comes along and says, "You can talk to the animals, and they can hear you and talk back to you," maybe that blows your whole mind about being safe in the universe. And also, look at agribusiness. If agribusiness acknowledged the fact that these animals feel fear, feel pain, have emotional lives, what would that say about the way we are treating them? Animal communication is very threatening to large segments of the commercial population.

It could undermine the whole economy. Are there times when it's hard to do this work?

It's hard when animals are dying. It's hard when people are not ready to let their animals go. Sometimes the animals will hold on for the sake of their people and endure whatever they need to endure.

And it's hard when I become emotionally attached. Even though I don't meet the people or the animals, there are some with whom, either because I've worked with them for a while, or because their stories are so touching to me, I become engaged in the whole process too.

You lose your boundaries.

I do. I try not to do it in the course of the communication, but I will get off the phone and cry. My husband has seen this many times. He used to say to me, "Maybe you shouldn't do this work because it makes you cry."

And I would say, "No, I'm crying because I'm grateful that I can do it. I'm crying because people have trusted me and the animals have allowed me to do it."

It's an honor.

It is an honor, and a privilege. The other part that's hard is when you're talking to someone who has unreasonable expectations of their animals. They want their animals to be absolutely perfect, to not be animals: You're in my house now, you can't move anything, you can't make any noise, you can't go on the couch, you can't go here, you can't go there. When I hear that I feel very bad because the animal is just being an animal. They want to be loved and they want to be close to their people and the people are the ones putting up the walls.

My biggest message, if there's a message, is that people need to know that animals are not lesser beings. That if people have chosen to be with animals and animals have chosen to be with them, there's so much they can learn from the animals: whether it's about being stoic, or living in the moment, or just loving life. I've even had animals who have given their lives for their people, who have willingly done this.

For example, a person once called me whose dog was partially paralyzed and needed assistance getting up and down steps. At one point the person fell on the dog, outside, and felt tremendously guilty about this, because it was the decline of the dog after that. The dog passed on shortly afterwards. But when we contacted the dog in spirit, the dog said that it was her privilege to be there to prevent the person from getting hurt.

To break the fall.

Yes! They do service for us all the time. They keep us grounded, they keep us in the moment, they lick our tears, they are there giving unconditional love.

How could someone who's interested in pursuing animal communication get started?

There are books and audio- and videotapes. There are close to a hundred and fifty communicators around the country now, some of whom, like myself, teach classes. A lot of people start by reading the books and listening to the tapes, and then go for further validation with classes. I would say that most people probably need to be able to get themselves centered and quiet. For me, that's happened through my meditation practice. A lot of people don't know what meditation is, but it's just sitting very quietly. You can pay attention to your breathing and just let your thoughts dissipate, if possible. It takes discipline, as you know. But you have to be able to get quiet in order to hear what the animals have to say. That's the first step.

Most people have had some experience, sometime during their life, of having a communication from an animal. I tell people to try to honor that experience and remember it and not think it was just a fluke. It's a real, valid phenomenon.

I believe that during communication there's an energy transference that takes place, which is why I say a protection prayer before each session. Once I was dealing with Lillian, a cat who had been bleeding from her mouth. I said, "Lillian, I want you to give me the feeling of what it was like," and suddenly I was retching and gagging. I could almost literally feel blood going down my throat. The next day I had a mouth full of canker sores under my tongue. They lasted about a week and a half. We later discovered that Lillian had cancer in her mouth. Eventually, that was what killed her.

After that experience I learned that I need to protect myself as much as I can. I actually ask for protection for everybody—for the person, for the animals, for anybody who lives in the household. I figure if the energy can go back and forth that easily we should be protected and not take on each other's stuff.

What are some of the most unusual conversations you've had with animals?

I spoke with a hedgehog once who was the first and last hedgehog I've spoken to. He lived in a cage in a library room.

I asked him what his life was like. What he said to me was, "I am primarily a solitary being." And I thought, Wow, that really says it all, doesn't it?

I have had animals tell me the exact verbiage that's used in a household, which I wouldn't have any way of knowing. My cousin had three cats. Two of them got along very well with each other, but the third cat wouldn't have anything to do with the other two. This had been going on for many years.

My cousin had me communicate with all of the cats. I asked each one in turn, "What's wrong? Why doesn't she get along with you?" The first cat said, "Well, you'll have to ask her." The second cat said, "Well, she has a real attitude." And the cat with the attitude, when I asked her, "Why don't you get along with the other two cats?" said, "Well, they're just *cats*." I thought that was pretty funny, coming from a cat. I said, "But you're a cat, too." And she said, "Well, *I* used to be a queen."

I had not met those cats. I did know their person, because she was my cousin. She said, "That's just how she acts. She acts like she's a queen!"

One day during the first year that I was communicating, one of my neighbors came to me and said, "There's a dog who lives here that you really have to talk to." The dog—I think it was an Akita—had only three legs. He had cancer, and his person had had a leg amputated rather than euthanize the dog. The dog refused to walk. The person would try to lure him with food, he would try to cajole him; the dog would take two steps and flop down, he just wouldn't go on. My neighbor said, "You really need to talk to this dog, he's very sad."

Was it a front leg or a back leg?

It was a back leg. I didn't know his person, but one day I was looking out of my window and across the street I saw the man walking the dog. The man was a saint. He was so patient with this dog. I thought, I've never done this without permission, but I'm sure that the man would want his dog to be happy. So I started communicating with the dog from the window.

I said, "My name is Gail, I'm a human being, and I speak to animals. I see that you're having some difficulty."

Before I heard any words, I was overcome with sadness. My eyes filled with tears. The dog said, "I'm so upset. I'm just embarrassed. I used to be able to run and walk, and look at me now. I can't do anything." I said to him, "Are you in great pain?" He said, "No." I said, "Can you understand that your person loves you so much that he could not bear to let you go? That this, he thought, was better—for them to take off your leg and to keep you with him as long as he possibly could. Can you understand that?"

And the dog said, "Well, I love him very much also." I said, "I know a dog from the shelter who had three legs"—and I did, she was a Rottweiler—"who used to bound up the steps"—

On the three legs?

On the three legs. She was in the shelter a long time. She was adopted, finally, by someone who lived in a third- or fourth-floor walkup. This dog would bound up the steps so fast that her person couldn't even keep up with her, but she would stop at each landing and wait. I said, "You know, if that dog can do this, I think you can do it, too. Life is a really beautiful thing, and you're alive now, and you're telling me you're not in pain. Will you just try to walk, and see what happens?" I said, "I hope some day I actually get to meet you."

And do you know, the dog got up, and he started pulling the man down the street. As far as I could see out of the window he pulled that man. And after that, he walked. He eventually died, because I guess the cancer was more than just in his leg, but he lived for over a year. I did get to meet him. I never told the man what I did because I didn't know how he would feel about it.

You didn't tell him?

No. To this day he doesn't know.

I'm a translator—

And that's exactly what I am, too.

And as a translator, I often feel that I can't just translate the words. Especially in the case of poetry, I have to have the experience, be in the writer's shoes and look through the writer's eyes and then describe what I see. So I'm wondering if it sometimes feels as if you don't know where the words are coming from. Does it feel as if the animal is actually speaking in words, or does it feel as if you're getting information that somehow comes through you and comes out words?

It's both. It depends on the animal. There are times when I just get the images and I'll say, "She's showing me this," or "I'm feeling that." There are other times when I'll hear, "Well, they're just *cats*."

I had another instance where a cat told me about another cat in the household. He said, "She thinks she's the queen of everything." And when I said that to the person on the telephone she burst out laughing. I asked her, "Why are you laughing?" She said, "Because over the door of the room where this cat lives there's a sign that says 'The Queen of Everything.' That's what we call her." Now, there's just no way that I could have known this.

You asked, Am I ever telepathic with the people? I will tell you very honestly, in the course of some communications I'm sure I'm picking up from the people as well as from the animals. But there are also communications where the people don't know what they're asking me about, and I will pick it up from the animal. For example, in cases of a lost animal, or an animal who is in spirit, the people may not know what happened to the animal. I will pick it up from the animal and then tell the person. The person may or may not know it at the time but can sometimes get it confirmed later.

Shortly after I started communicating with animals in spirit, a friend of mine who was recently divorced called me up. She said, "My mother-in-law's dog passed away last fall. I just heard about it. I feel terrible." She had lived in the house with the dog. He was a seven- or eight-year-old Labrador retriever. "Could you ask him what happened? And tell him I'm very sorry." She also had some questions about the daily life of the dog in the household. I said, "I'll try. I don't know what I'll get." This was back in the days before I did sessions with the person actually on the telephone with me.

So I sat down by myself, tuned in, and asked to speak with this dog. The first thing I heard was, "I wasn't seven or eight years old when I left my body. I was ten." He gave me details about daily life in the household. I said, "Why did you have to leave your body?" He said, "I couldn't eat, I couldn't defecate. After a few days they had to release me." I wrote it all down and I called my friend back. I said, "This is what the dog told me. I have no idea if any of it's true. If you can confirm any of it, I would appreciate it."

A couple of weeks later I got a message on my voice mail. In this quavery voice my friend said, "You were right, he was ten. They didn't know if he had intestinal cancer or a blockage of some sort. The problem was in his gastrointestinal tract. They tried to help him for a few days and they finally had to release him." Everything I got about the family life was accurate. Here was a case where the client did not know what she was asking me to find out. It came through. And the hair on the back of my neck went up. That was one of the more dramatic cases.

The first time I spoke with you was shortly after I had to put Daisy in the basement because she could no longer climb the stairs. I asked you to ask Daisy, How can I make your life better? You said she showed you a picture of me living in the basement and sleeping on a couch that was down there. That really impressed me because you had no way of knowing that there was a couch in my basement. Daisy told you that. How many basements are fully furnished with beds and couches?

Some are. Somebody could say, Well, that was a good guess on her part. And the old me would have said that, too. They say that God is in the details. Sometimes the things that the animals tell me are so little, but they're so appropriate.

One dog that I was tuning in to kept telling me about his toys. He said, "You just can't believe how wonderful my toys are. I have so many toys." A lot of people would say, Well, most dogs have a lot of toys. But this dog had a huge box of toys and it was his pride and joy.

I recently worked with a dog in spirit. The first thing the dog said to me was, "Tell her I'm running, I can run." I thought maybe the dog had been crippled. I figured out that the dog had been hit by a car, because I could feel all the chest trauma. The woman said, "Since she's passed on, I've been saying to her in spirit, 'Just run like the wind.'" And that was the first thing the dog said. Not even "Hello," but "Tell her I can run."

So sometimes you pick information up psychically from the person?

Yes, I'm sure.

It even occurred to my skeptical side that that might have been the case with the bed in the basement.

What I think about that is that the animals are telepathically open to you at all times. If you have something in your consciousness, they have it in their consciousness. So it almost doesn't matter in some cases whether you're getting it from the person or the animal. It's the same consciousness.

Because we're all one. And that's the whole message of this work. We're all connected.

Right. But there are times when people don't know what they ask me to find out, and I still get the answers.

I don't do many lost animals anymore, and only if the people and animals are known to me. Sometimes lost animals don't want to be found, and then I'm right in the middle of a situation where the person is saying, "Locate my animal," and the animal is saying, "I can't go back there any more," or "I don't want to stay in my body."

A woman called me from Florida who had an elderly Yorkshire terrier whom I knew. The dog had very bad cataracts, and I think she was even deaf at this point. Her person had moved to a new home. The dog got out, and the woman was beside herself. Could I help her, because the poor dog can't see, can't hear, and it's cold. So I said, "All right, let me try."

I sat down, meditated, and asked the dog to show me where she was. I saw a room on a ground floor with pastel colors; the dog was inside. I could see a woman who had either blond or light gray hair and I could feel that she was very kind. She was wearing a light pink sweater set. The dog was perfectly fine, she was being taken care of. I could see tall grass outside.

I called my client back and said, "This is what I'm getting. Maybe it's just my projection because I don't want anything bad to happen to this dog. I think she's all right, she doesn't seem to be frightened. This woman is very happy to have your dog. She is about your age, she's very kind. Do all of the usual things: put up posters, talk to people, look around. Please call and let me know what is happening."

My client called me back the next day. She said, "Everything you said was true, right down to the colors on the walls. She's a woman about my age, she has light hair, she was thrilled to have my dog. She saw the posters, and she got the dog back to me."

Another communicator, Ginny Debbink, and I worked on a bird who got lost here in New York City. This was the bird of a friend of mine, a little monk parrot named Beeper, a very funny bird. Beeper was free-flying in the house and somehow got out the window. He was out on the terrace when the husband came home. Beeper flew to the man a couple of times, and then he flew away, down the street. My friend called me up, distraught. She said, "He's a domesticated bird, he's out in the city, I don't know what to do." I said, "All right, let me see what I can do."

I began to tune in to Beeper. I could tell the direction in which he had gone. I said, "There's a high-rise apartment building with light-colored brick. Up on the top there are plants, a penthouse area. He's up there. What I think you'd better do is take a picture of your bird with you and go and knock on doors. He's three or four blocks away from you." She went out and knocked on doors, and people were very generous. They let her go up on the tops of their buildings. I couldn't believe it, because in New York City you don't know what kind of a reaction you're going to get.

Anyway, this horrible storm came up. It was pouring rain sideways, it was thundering and lightning, and the wind was blowing, and yet Beeper was indicating that he was perfectly fine. I said, "I don't know what's going on, but he's okay." Again, I thought that maybe I was just projecting this because I knew Beeper and I didn't want anything bad to happen to him.

So the next morning I said, "Look, I don't do lost animal work so much. Why don't you call Ginny, because she's really good with lost animals." Ginny tuned in and got the same kind of thing, but the additional piece of the puzzle she had was that the bird could see the river. It gave my friend another direction to look in.

East River or Hudson?

Hudson River; this was on the West Side. So my friend went out again and knocked on one particular door, and she told the doorman, "I've lost my parrot. We think he might be

in the penthouse of one of these buildings." He said, "Is your parrot really a loud bird?" She said, "Yes." He said, "Just a minute." He called upstairs, and the bird had flown into the penthouse apartment.

The people in the penthouse had handled birds before. They took Beeper in, they took him to the vet, they put him in a cage. That's why he wasn't out in the storm the night before. My friends got Beeper back. It was amazing.

With animal communication, every session is different. It's partly because of the animals' way of perceiving. I worked with a dog who, when I asked her, "What's your life like?"

showed me layers of light and colors. It was almost like looking at an Impressionist painting. I was seeing green plants and all this light coming through the window and the little girls who lived in the family. The dog said, "There's a time every day when we have to be very quiet because the girls are busy."

I told all this to the person, who happened to be sitting there with me, and all of it was confirmed. There were a lot of plants in the house, there was good light coming through the windows, and the girls have to do their homework in the afternoon when they come home from school.

Most times, I just ask the question and I get a picture, or I hear the words, and then we go on from there; but this time the visual images just kept on going and going and going.

Do you remember what breed of dog it was?

Yes, I do, she was a Chinese crested.

So you never know what to expect when you begin a communication.

No, I never do. And that's part of the excitement of it. But it can also be terrifying, because you just don't know where you're going with these things. Somebody once told me that when you have these psychic abilities, you have to have the courage to jump off the cliff. Every once in a while I get to a place where I don't think I can do it. I can, but I make it harder for myself when I get to that place. I have had animals say to me, "Stop being fearful about this. This is what you were put on this planet to do; just do it." And I would

say, "But I don't know enough." I once had a llama say, "Your whole life is learning, and after you leave your body, then you do a whole other kind of learning." I just try to honor what the animals keep telling me.

When I give a talk on animal communication to a club or a group, I look down in the audience, and invariably there's someone sitting there with tears running down their cheeks. Because that has been their experience, but everybody told them they were crazy; or they were too afraid to express it because *they* thought they were crazy. When I teach, I am more or less empowering people to remember and to relearn what they already really know.

The irony in this is that no one can teach you how to be an animal communicator. I can tell you that this is a real thing and that you are capable of doing it. I can give you exercises to do, and I can tell you stories about things that have happened to me. Then you go on your own journey. Everybody's on their own journey with it. That's how I feel about it. It's my spiritual journey.

Epilogue

We celebrate birthdays, we need to celebrate deathdays as well. It is easy to celebrate a birth: a new life, a clean slate, all innocence and purity and trust. It is less easy, but just as important, to celebrate a death: a mission accomplished, a moment of completion, perspective, and wisdom. At the death of a loved one their whole life lies before us in all its fullness and variety, a rich tapestry of experience with the spirit rising from it, the spirit that entered the baby and is now free.

My father died the day after Thanksgiving, so when the trees are bare and the cold winds of winter have begun to blow and our family gathers to celebrate the holiday, my father's spirit is a palpable presence, as real as the bleak November landscape and the laden board.

Mother died in early May, so at that time of year when the Earth has awakened from her long winter's nap, the flowering trees of northeast America will always call up for me her gardener's spirit and her love of beauty, and I look at the blossoms through her eyes.

Daisy died in mid-August, in the heart of midsummer, a time when the trees are so heavy with foliage that I can no longer see my neighbor's houses and the insect chorus booms all night.

In late July of the following summer, right on schedule, the katydids struck up their nightly concert, reminding me of the late-night walks Daisy and I took the year before. A few days before the anniversary of her death a cricket took up residence in my bathroom, and every evening she serenaded me, just as another cricket had done the year before, right after Daisy died.

As the first anniversary of her death drew near, I knew I had to mark it in some way. Through the animal communicators I had had six conversations with my dog. What better way to honor her than by speaking with her again? The book was almost ready to go into production, but it was still not too late to add a few words if Daisy had a final message for our readers. And I was curious to know where she was, how she was doing, and whether she had any plans to reincarnate.

Epilogue

I called Gail De Sciose and made a date.

The day before our appointment I spent the afternoon with a friend, her daughter-in-law, her grandson, and their dog, a female beagle-Border collie mix. Brownie's delicacy, beauty, and enthusiasm for life reminded me so much of Daisy that I couldn't stop petting her. I realized again how much I missed having a dog in my life. At the same time I knew I wasn't ready to take on the responsibility of a puppy. I almost felt guilty asking Daisy her plans when mine were so uncertain.

At the appointed time I called Gail, and as she tuned in to Daisy, there was a long pause. Had she reincarnated? Was she busy leading another life? I didn't think so, because it felt like she had been helping me with our book.

SEVENTH CONVERSATION WITH DAISY, AUGUST 17, 1999

> At last Gail spoke. "Daisy says your hearts are so linked there is no separation between you. She says, *Helen speaks to me all the time.*"
>
> "It's true! I'm always talking to her. I'm so glad she can hear me! Can you ask her where she is?"
>
> "This is very unusual. I don't often get such a detailed picture. . . .
>
> "She's showing me a river valley. She is following the river and exploring along the bank and into caves. There are two spirit beings with her, a man and a woman. Sometimes they send her on alone,

but she always comes back. She is having a great time."

"I think of her all the time, especially when I am walking in the places where we used to walk and when I hear the katydids singing at night like they did last year just before she died. I always wonder, can she hear them, too?"

"She says she hears them, and she also hears frogs and crickets."

"That's right, there's a cricket in the house again. She sings to me every night. Please tell Daisy it is exactly a year since she made her transition, and ask her if she has any plans to take another body."

"Daisy says she knows that time has passed, but she has no sense of how much time. It feels to her like only an instant since she was with you. She says, *I will come back, but not right away.* She says—what a being!—she says, *I would be an interference if I came back right now. Helen needs to be very focused on her purpose.* She says she is doing a lot of good for you by being your muse."

"That is amazing. She knows I'm not ready, and she's willing to wait. I have kind of a strange question. That photograph I took of her back in Connecticut, the one I'm going to put on the cover of the book, is so extraordinary. How often does an animal look right into the lens of the camera with such intensity? When I took that picture, did Daisy know I was going to write a book about her?"

"Daisy says you took many pictures of her, not all as direct as that one. She didn't know what would be done with this picture, but she felt it was important for people to look right in her eyes, because when people look into her eyes, they can see her soul."

"Does she have a last message for our readers?"

"She says, *Just tell everyone that it's all about love. When you have an animal, see how they are able to love, and let that be a model for you. The greatest gift animals can give you is their love. People need to love themselves and each other in that way.*"

"Please tell her how grateful I am to her, how much I love her, and how wonderful it is working with her."

"She says, *Do you see how much help I can be by staying in spirit right now?*"

"She is right, I need her to be in spirit. I guess I have to say goodbye, although I know she's always with me. Please send her a big hug. Tell her, I'm scratching her belly and wiggling her ears the way I used to do when she was in her body."

Gail laughed. "She says, *It's enough to make me want to come right back!* She says she is very proud of you. Do you have a braided rug?"

"Yes, my grandmother made it."

"She's showing me a braided rug. She says when you're up late at night working on the book, she is right beside you, on that rug."

"Tell her we're a good team."

A Word to the Skeptical Reader

Conversations with a dog? I would be the last to deny that there is much in this little book to stretch the credulity of the skeptic.

What is a skeptic? The word comes from the Greek *skopos*, watcher (ironically, it is related to the word *horoscope,* a map produced by watching the sky). Originally, it referred to a school of philosophers who believed that absolute knowledge was impossible, but that observation was more reliable than human reason.

With the decline in religious belief, the persistence of the eighteenth-century faith in reason, and the concomitant modern aversion to the so-called irrational, the word has come to have a slightly different connotation that no longer conveys the old mistrust of reason. In any event, I use the word "skeptic" in its modern sense as almost synonymous with "rational materialist": one who believes that only that which can be perceived with the five senses, only that which mainstream science can observe, measure, and make predictions about, is real.

If you have read this far—indeed, if you have picked up this book at all—you may be a member of that perennially endangered species, the open-minded skeptic: a skeptic in the original sense of one who doubts but is willing to examine the evidence. A skeptic in this sense may say "I don't believe this" but will not therefore conclude "so it's not worth investigating."

My more skeptical friends who have read this book in manuscript may have praised its "sensitivity," but on the question of animal communication, most have maintained a polite but eloquent silence. I don't blame them. A scientist's daughter, I was a skeptic until well into my thirties.

But things started happening that made cracks in the rational materialist belief system I was raised in, and the light those cracks let in made life a little more interesting. I started seeing connections I hadn't noticed before—synchronicities, if you will—and at a certain point they began adding up, began tipping the balance in favor of a universe that was more than mere physical matter, more than a series of improbable accidents: a universe with meaning and intelligence.

I think I almost made a conscious decision to live "as if": as if the universe had meaning and purpose, as if synchronicity was not just a fancy name for coincidence, as if there was a hidden design that included everything and everyone, even though I couldn't always see it. I started to have faith: not in God, necessarily, but in what I can only describe as the power of faith itself—the power of faith to heal and to enlighten, to improve the quality of life, maybe even to explain it. And indeed, I found that life lived with faith, even if it was just faith-in-the-power-of-faith, went better.

I felt more connected to the other lives around me, to the Earth, to the universe. I became less depressed, more loving. I started having more fun. And this in turn suggested something about the nature of the universe. It seemed a confirmation that I was somehow on the right track, that the universe had a kind of internal consistency and unity that was not only reasonable but aesthetically pleasing. My working theory of a meaningful universe had that quality of elegance and simplicity that tells the mathematician the proof is right. It was a sort of experiment in belief, and looking back, I feel it's been a success.

In *Pack of Two: The Intricate Bond Between People and Dogs*, Caroline Knapp gives animal communication short shrift. A thorough researcher, she does consult a couple of so-called animal psychics, and does not doubt their sincerity, but "interspecies telepathic communication" is not her cup of tea. She acknowledges that her dog Lucille has an uncanny ability to sense her needs, wants, and feelings, but she insists that she can't know what is going on inside her dog's mind—a kind of reverse speciesism I find quite remarkable. The truth, she admits, is that she really prefers *not* to know, that she wants her dog to remain mysterious.

This is the hidden romanticism of the rational materialist position. Having ruled out *a priori* the mystery of interspecies communication, these modern-day skeptics retain a kind of nostalgia for mystery but prefer not to examine it too closely.

Knapp's aversion to animal communication, her need for her dog to remain a mystery, reminds me of the discomfort my father, Warren Weaver, felt back in the 1930s and 40s when J.B. Rhine, the pioneering researcher in extrasensory perception, started piling up statistical evidence for ESP. Dad said the whole subject made him so nervous he didn't like to think about it, but he knew that wasn't the right attitude. In spite of his nervousness, he was later instrumental in getting funding for Rhine's Parapsychology Laboratory at Duke University. Dad was a skeptic in the classical sense. Not many scientists have that kind of integrity.

This discomfort with phenomena like animal communication, ESP, and so on, that seem to undermine the logical foundations of science, masks a deep fear. For what if these things were real? Then where would we be? We would be in a universe where everything is connected, everything has a purpose, and everything is alive. That *is* scary, because there's no place to hide. No ivory tower of scientific immunity; no separation between human observer and experimental subject.

The history of science has seen one long series of blows to the human ego. Copernicus maintained that our Earth is not the center of the universe; Darwin found that we share a common ancestor with the apes; Freud's work on the unconscious implied that we are not even the masters of our own minds. A growing body of evidence suggests that we may not be the only intelligent beings in the universe. If animal communication is real, then we are not even the only intelligent beings on our own planet. All of these ideas have initially been rejected as absurd. We need to remember that the mere fact that an idea is threatening to the mind does not prove it invalid.

The idea that the universe has design and consciousness is not inherently any less probable than the idea that it is random. The clinging to coincidence and accident as the sole explanation for phenomena that suggest a connecting link (however mysterious) seems just as irrational, if not more so, than the acceptance of the possibility of connections we do not yet understand.

Whatever the explanation of these mysteries—and as my father's daughter I am confident we are making progress and will track them down in the end—we humans need to remember that we are connected to the animals. We need to know that communication with them is not only possible but natural, and not only natural but essential. Our separation from the rest of nature, our arrogant assumption that we are somehow in charge, is laying waste the planet and may well lead to our own extinction unless we can allow ourselves to rediscover this connection, and to start listening.

Resources

SUGGESTED BOOKS AND TAPES

On Animals and Animal Communication

As this book goes to press on the threshold of a new millennium, it seems that scarcely a week goes by without the publication of another book about animals. And whether the authors are writing from a scientific background or from a New Age perspective (or both, as occasionally happens), these books tend to have a common theme. They tend to agree that the non-human animals are more intelligent and more emotionally complex than we have been accustomed to believe—in other words, that they are more like us.

The sin of anthropomorphic thinking—the ascribing of human characteristics to animals—is being called into question. The grim legacy of René Descartes, whose belief that animals are unfeeling machines enabled him to dissect live, unanaesthetized dogs and ignore their howls of pain, is gradually giving way to the less mechanistic views of Charles Darwin, whose focus on the similarities between species led him to the revolutionary notion that animals have emotional lives. In a dramatic indication of changing attitudes, the Harvard Law School is now offering its first-ever course on animal law.

The idea that animals are intelligent and sensitive individuals who have much to teach us about ourselves and legal rights of their own is truly an idea whose time has come.

As a consequence, keeping the following list of books on animals and animal communication up to date has been an impossible, if agreeable, task; and by the time this book is published, the list will of necessity be incomplete. Should *The Daisy Sutra* see further editions, I will attempt to keep abreast of the literature. In the meantime, all honor to these courageous pioneers!

Resources

Adams, Janine. *You Can Talk to Your Animals: Animal Communicators Tell You How*. Foster City, CA: Howell Book House, 2000.

Boone, J. Allen. *Kinship with All Life*. New York: HarperCollins Publishers, 1954.

Curtis, Anita. *Animal Wisdom: Communications with Animals. 1996.* Available from the author at P. O. Box 182, Gilbertsville, PA 19525, (610) 327-3820.

Curtis, Anita. *How to Hear the Animals.* 1998. Available from the author at P. O. Box 182, Gilbertsville, PA 19525, (610) 327-3820.

Fitzpatrick, Sonya, with Smith, Patricia Burkhart. *What the Animals Tell Me: Developing Your Innate Telepathic Skills to Understand and Communicate with your Pets.* New York: Hyperion, 1997.

Goodall, Jane, with Berman, Phillip. *Reason for Hope: A Spiritual Journey.* New York: Warner Books, 1999.

Hiby, Lydia, with Weintraub, Bonnie S. *Conversations with Animals: Cherished Messages and Memories as told by an Animal Communicator.* Troutdale, Oregon: NewSage Press, 1998. Available from NewSage Press, P. O. Box 607, Troutdale, OR 97060-0607, (503) 695-2211.

Kowalski, Gary. *The Souls of Animals.* Walpole, NH: Stillpoint Publishing, 1991.

Lauck, Joanne Elizabeth. *The Voice of the Infinite in the Small: Revisioning the Insect-Human Connection.* Mill Spring, NC: Swan Raven & Co., 1998. Available from Blue Water Publishing, Inc., P. O. Box 190, Mill Spring, NC 28756, (800) 366-0264.

Masson, Jeffrey Moussaieff. *Dogs Never Lie About Love.* New York: Three Rivers Press, 1998.

Masson, Jeffrey Moussaieff, and McCarthy, Susan. *When Elephants Weep: The Emotional Lives of Animals.* New York: Delacorte Press, 1995.

McElroy, Susan Chernak. *Animals as Teachers & Healers: True Stories and Reflections.* New York: Ballantine Books, 1997.

Mowat, Farley. *Never Cry Wolf.* Boston: Little, Brown and Company, 1963.

Myers, Arthur. *Communicating with Animals: The Spiritual Connection Between People and Animals.* Chicago: Contemporary Publishing Company, 1997.

Page, George. *Inside the Animal Mind: A Groundbreaking Exploration of Animal Intelligence.* New York: Doubleday, 1999.

Resources

The Psychic Connection: Exploring the Spiritual Link Between People and Animals (videotape). Palm Springs, California: The Entertainment Group, 1999. Available from Anita Curtis at P. O. Box 182, Gilbertsville, PA 19525, (610) 327-3820.

Sheldrake, Rupert. *Dogs That Know When Their Owners Are Coming Home, and Other Unexplained Powers of Animals.* New York: Crown Publishing Group, 1999.

Singer, Peter. *Animal Liberation.* New York: The New York Review of Books, 1975

Smith, Penelope. *Animal Death: A Spiritual Journey* (audiotape). Available from Upper Access at (800) 356-9315 or on the web at upperaccess.com.

Smith, Penelope. *Animal Talk: Interspecies Telepathic Communication*. Hillsboro, OR: Beyond Words Publishing, 1999. Available from (800) 356-9315 or on the web at upperaccess.com. (Note: This is a revised and expanded edition of the original *Animal Talk,* Pegasus Publications, 1982.)

Smith, Penelope. *Interspecies Telepathic Connection Tape Series* (six one-hour audiotapes). Available from Upper Access at (800) 356-9315 or on the web at upperaccess.com.

Smith, Penelope. *Species Link* (quarterly journal). Available from Pegasus Publications at P. O. Box 1060, Point Reyes, CA 94956, (415) 663-1247.

Smith, Penelope. *Telepathic Communication with Animals* (videotape). Available from Upper Access at (800) 356-9315 or on the web at upperaccess.com.

Smith, Penelope. *When Animals Speak: Advanced Interspecies Communication*. Hillsboro, OR: Beyond Words Publishing, 1999. Available at: (800) 356-9315 or on the web at upperaccess.com. (Note: This is a shorter version of *Animals: Our Return to Wholeness*, which was published by Pegasus Publications in 1993.)

Summers, Patty. *Talking with the Animals*. Charlottesville, VA: Hampton Roads Publishing Company, Inc., 1998.

Thomas, Elizabeth Marshall. *The Hidden Life of Dogs*. New York: Houghton Mifflin Company, 1993.

Tobias, Michael, and Solisti-Mattelon, Kate, eds. *Kinship with the Animals*. Hillsboro, OR: Beyond Words Publishing, Inc., 1998.

Wagner, Teresa. *Legacies of Love: A Gentle Guide to Healing from the Loss of Your Animal Loved One* (two audiotapes). Available from Upper Access at (800) 356-9315.

Wise, Steven M. *Rattling the Cage: Toward Legal Rights for Animals*. New York: Perseus Publishing, 1999.

Wright, Machaelle Small. *Behaving as if the God in all Life Mattered.* Jeffersonton, VA: Perelandra Ltd., 1987. Available from Perelandra, Ltd. at Box 3603, Warrenton, VA 22186.

On Meditation

Since most animal communicators practice some form of meditation and many of them have come to this work through meditation, I have included a few books on this subject from several spiritual and religious traditions. Those starred are recommended for beginners.

Bair, Puran. *Living from the Heart: Heart Rhythm Meditation for Energy, Clarity, Peace, Joy, and Inner Power*. New York: Three Rivers Press, 1998.

Bodian, Stephan. *Meditation for Dummies.** Foster City, CA: IDG Books Worldwide, Inc., 1999.

Goldsmith, Joel. *The Art of Meditation.* San Francisco: Harper & Row, 1990.

Kaplan, Aryeh. *Jewish Meditation: A Practical Guide.* New York: Schocken Books, 1985.

Le Shan, Lawrence. *How to Meditate: A Guide to Self-Discovery.** New York: Bantam Books, 1974.

Muktananda, Swami. *Meditate: Happiness Lies Within You.* South Fallsburg, New York: Syda Foundation, 1999.

Nhat Hanh, Thich. *The Miracle of Mindfulness: A Manual on Meditation.** Boston: Beacon Press, 1987.

Smith, Jean, ed. *Breath Sweeps Mind: A First Guide to Meditation Practice.** New York: Riverhead Books, 1998.

Strand, Clark. *The Wooden Bowl.** New York: Hyperion, 1998.

St. Ruth, Diana. *Sitting: A Guide to Buddhist Meditation.** New York: Penguin Arkana, 1998.

Resources

On Life After Death and Reincarnation

Quite a few animal communicators find themselves dealing with issues of loss, life after death, and reincarnation. Here again, those titles starred will be more accessible to readers unfamiliar with these subjects.

Bernstein, Morey. *The Search for Bridey Murphy.** New York: Doubleday, 1989 (out of print).

Brinkley, Dannion. *Saved by the Light.** New York: Harper Collins, 1995.

Eadie, Betty J. *Embraced by the Light.** New York: Bantam Books, 1994.

MacLaine, Shirley. *Out on a Limb.** New York: Bantam Books, 1986.

Moody, Raymond A., Jr., M.D.. *Life After Life.** New York: Bantam, 1976.

Roberts, Jane. *Seth Speaks: The Eternal Validity of the Soul.* San Rafael, CA: Amber-Allen Publishing, 1994.

Stevenson, Ian. *Children Who Remember Previous Lives: A Question of Reincarnation.* Charlottesville, VA: University of Virginia Press, 1987.

Weiss, Brian L. *Many Lives, Many Masters.* New York: Simon and Schuster, 1988.

Woolger, Roger. *Other Lives, Other Selves: A Jungian Psychotherapist Discovers Past Lives*. New York: Bantam Books, 1988.

Yarbro, Chelsea Quinn. *Messages from Michael.** New York: Berkley Books, 1983.

ANIMAL COMMUNICATORS

The following list is by no means exhaustive. No doubt there are many fine animal communicators whose names do not appear here. This list represents only people I have consulted myself, or their referrals. Those whose names are starred periodically offer introductory classes or workshops in animal communication.

Sharon Callahan*
Anaflora Flower Essence Therapy for Animals
P. O. Box 1056
Mt. Shasta, CA 96067
(530) 926-6424 (phone)
(530) 926-1245 (fax)
anaflora@snowcrest.net
www.anaflora.com

Resources

Anita Curtis*
P. O. Box 182
Gilbertsville, PA 19525-0182
(610) 327-3820 (phone)
(610) 970-2696 (fax)
amicom@aol.com
www.anitacurtis.com

Virginia Debbink
51 Schooley's Mountain Road
Long Valley, NJ 07853
(908) 876-9442
debbink@interpow.net
vdebbink@drew.edu

Gail De Sciose*
1623 Third Avenue, 5K West
New York, NY 10128
(212) 831-4666

Sue Goodrich
1205 Bear Valley Parkway
Escondido, CA 92027
(760) 480-2474
sgoodrich1@earthlink.net

Carol Gurney*
3715 North Cornell Road
Agoura, CA 91301
(818) 597-1154
cgurney@earthlink.net
www.animalcommunicator.net

Jane Hallander
Novato, CA
(415) 899-9998
jing@ix.netcom.com
www.netcom.com/~jing

Dawn Hayman*
Spring Farm Cares
3364 State Route 12
Clinton, NY 13323
(315) 737-9339
www.springfarmcares.org

Morgine Jurdan*
1135 Yale Bridge Road
Amboy, WA 98601
(360) 247-7284
Toll Free for consultations only:
(877) 667-4463
morgine@hotmail.com

Sam Louie*
P. O. Box 2051-A
Berkeley, CA 94702
(510) 644-1583
sl@samlouie.com
www.samlouie.com

Raphaela Pope*
P. O. Box 2062
Davis, CA 95617
(530) 758-6111

Jeri Ryan*
P. O. Box 10166
Oakland, CA 94610
(510) 569-6123

Marlene Sandler*
P. O. Box 476
Warrington, PA 18976
(215) 491-0707

Joanna Seere*
P. O. Box 1017
Planetarium Station, NY 10024
(212) 595-4336

Penelope Smith*
Pegasus Publications
P. O. Box 1060
Point Reyes, CA 94956
(415) 663-1247
www.animaltalk.net

Kate Solisti-Mattelon and Patrice Mattelon*
13 Bluebell Court
Santa Fe, NM 87505
(505) 466-6958
solmat@earthlink.net
http://home.earthlink.net/~solmat

Teresa Wagner*
Monterey, CA
(916) 454-0308
legaciesoflove@aol.com
teresa@animalsinourhearts.com

About the Author

As a professional literary translator, Helen Weaver has rendered some fifty books from the French. Her translation of *The Selected Writings of Antonin Artaud* was nominated for a National Book Award. She is co-author and general editor of *The Larousse Encyclopedia of Astrology*. She lives in Woodstock, New York, where she is at work on her next book, a memoir of the fifties.

About the Illustrator

Alan McKnight has illustrated more than 200 books, including the collector's editions of Louis L'Amour's western fiction and Bob Berman's *The Secrets of the Night Sky.* He lives in Victor, Idaho, where he hikes and paints watercolors in the wild lands of the Rocky Mountains, illustrates books and articles, and occasionally teaches.